WITHDRAWN

Aflatoxins and Human Health

Author

Ivana Dvořáčková, M.D.

Department of Pathology
Faculty Hospital
Charles University
Hradec Králové, Czechoslovakia

CRC Press, Inc.
Boca Raton, Florida

THE PENNSYLVANIA STATE UNIVERSITY
COMMONWEALTH CAMPUS LIBRARIES
YORK

Library of Congress Cataloging-in-Publication Data

Dvořáčková, Ivana, 1929—
    Aflatoxins and human health / author, Ivana Dvořáčková.
      p. cm.
    Includes bibliographies and index.
    ISBN 0-8493-4628-2
    1. Aflatoxins—Toxicology. 2. Food poisoning. I. Title.
    RA1242.A344D86   1990   89-37491
    615.9′5292—dc20   CIP

This book represents information obtained from authentic and highly regarded sources. Reprinted material is quoted with permission, and sources are indicated. A wide variety of references are listed. Every reasonable effort has been made to give reliable data and information, but the author and the publisher cannot assume responsibility for the validity of all materials or for the consequences of their use.

All rights reserved. This book, or any parts thereof, may not be reproduced in any form without written consent from the publisher.

Direct all inquiries to CRC Press, Inc., 2000 Corporate Blvd., N.W., Boca Raton, Florida, 33431.

© 1990 by CRC Press, Inc.

International Standard Book Number 0-8493-4628-2

Library of Congress Card Number 89-37491
Printed in the United States

DEDICATION

To the memory of my auntie Jarmila for all her love, and to the memory of Dr. A. D. Campbell, FDA, Washington, D.C., who did so much pioneer work in the field of mycotoxins and encouraged me to write this book many years ago.

My thanks belong to my husband for his support and patience.

PREFACE

The modern human population is constantly exposed to environmental hazards such as chemical dumps, polychlorinated biphenyls, aromatic amines, insecticides, herbicides, and a host of others, which may be dangerous to man due to their toxic and carcinogenic effects. Besides the industrial pollutants, many toxic and carcinogenic substances of natural origin that may be involved in human diseases have been discovered in the last few years. Among them the toxic mold metabolites — mycotoxins — are of particular importance.

Less than 25 years ago, the mycotoxicoses were called "the neglected diseases". The situation, however, has changed drastically since then, and they are now receiving worldwide attention. Isolation and characterization of mycotoxins that may occur in food and feed products are now major objectives of much of the current research.

Almost all plant products can serve as substrates for fungal growth and subsequent mycotoxin formation, thus providing the potential for direct contamination of man's food. When farm animals raised for food production ingest feed contaminated with mycotoxins, not only might it have a direct toxic effect on the animals, but there might also be a carryover of the toxins into man's food, thus creating a further way of human exposure to mycotoxins. Furthermore, occupational exposure may occur through other media, such as air.

Among the mycotoxins, particularly the toxic metabolites, the molds *Aspergillus flavus* and *parasiticus* — aflatoxins — have been intensively studied and shown to be associated with effects on human health.

The available human data relating to aflatoxins are mostly based on the epidemiological studies, while the pathological studies relating to morphological changes due to aflatoxin effects on man are very few.

Pathology, in common with all branches of science, is forever changing, as new concepts of disease and new techniques are developed and applied. Being largely concerned with descriptions of morphological changes in diseased organs, it is becoming increasingly concerned with efforts to understand how such changes occur. Thus the role of a pathologist is to insist on environmental histories to actively clear up the etiology of unfamiliar diseases and cancer etiology. The spirit of inquiry necessitates a new interdisciplinary approach. The development of immunology and enzyme chemistry and the application of new physical-chemical methods are rapidly changing some of the older concepts of diseases and helping to elucidate etiology.

For many years, Dr. A. D. Campbell, who held positions of responsibility in the Food and Drug Administration, Washington, D.C. and the Association of Official Analytical Chemists (AOAC), concerned himself with the presence of aflatoxins in food products relating to health risks of the human population and had a profound influence on the speed with which new research developments were disseminated and used.

I would like to pay tribute to the memory of Dr. Campbell for his help and encouragement to write this book, the aim of which is to bring together up-to-date available data concerned with health risks due to aflatoxins in humans.

Ivana Dvořáčková

THE AUTHOR

Ivana Dvořáčková, M.D., is a scientific specialist in the Department of Pathology of the Faculty Hospital at the Charles University in Hradec Králové, Czechoslovakia. She obtained her M.D. degree in 1953 from Charles University in Prague. She is an external worker of the Czechoslovak Academy of Sciences and received her C.Sc. degree from the French National Academy of Medicine in 1980.

Dr. Dvořáčková is a member of the Purkyne's Society, European Pathological Society, National Reye's Syndrome Foundation, Mycotoxin Working Group IUPAC, Association A. Tessier, and a member of the Scientific Advisory Board of the Journal of Microbiology, Aliments, and Nutrition.

She has been the recipient of research grants from WHO (1979) and from INSERM (1981).

She has been the recipient of the prize Pierre et Céline l'Hermite and of the diploma of the Laureate of the French National Academy (1985).

Dr. Dvořáčková has presented over 20 invited lectures at international meetings and over 80 lectures at national meetings. She has published more than 70 research papers and monographs. Her research interests include the mycotoxins and their harmful effects on the human health.

TABLE OF CONTENTS

Chapter 1
Mycotoxins and Human Diseases 1
I. History of Diseases Due to Mycotoxins in Man 1
II. Aflatoxins 2
    A. Occurrence and Sources of Aflatoxins 3
    B. Methods of Analysis 4
    C. Toxicity 4
    D. Carcinogenicity 5
    E. Teratogenic Activity 5
    F. Mutagenicity 6
    G. Immunological Effects 6
    H. Metabolic Activation of Aflatoxin 7
        1. Aflatoxin B_2a and G_2a 7
        2. Aflatoxicol 7
        3. Aflatoxin P_1 8
        4. Aflatoxin Q_1 8
        5. Highly Active Forms of AFB_1 8
    I. Biochemical Effects of Aflatoxin 8
    J. Factors Modifying the Effects of Aflatoxins 9
        1. Sex Differences 9
        2. Endocrine Status 9
        3. Nutritional Factors 9
III. Effects of Aflatoxin on Man 10
    A. Exposure Conditions 10
    B. Acute Effects of Aflatoxin 10
    C. Carcinogenic Effect of Aflatoxin 12
References 13

Chapter 2
Aflatoxin and Reye's Syndrome 21
I. Introduction 21
II. History of Reye's Syndrome 21
    A. Clinical Symptoms 21
    B. Laboratory Data 21
    C. Therapy 22
    D. Morphological Features 23
        1. Gross Pathological Changes 23
        2. Histological Changes 24
        3. Ultrastructural Changes 24
    E. Pathogenesis 25
    F. Epidemiological Study in the U.S. 27
        1. Antecedent Illnesses and Seasonal Distribution of RS 28
        2. Age 28
        3. Sex and Race 28
        4. Mortality 29
        5. Rural Clustering 29
    G. Epidemiological Studies Outside the U.S. 29
        1. Reye's Syndrome in Great Britain 29
        2. Reye's Syndrome in Asia 30

  H. Etiology 30
    1. Viral Infection 30
    2. Genetic Factors 31
    3. Intrinsic and Extrinsic Toxins 32
    4. Salicylates 32
    5. Environmental Toxins 33
    6. Aflatoxins 33
    7. Interaction of Viruses and Toxins 34
References 35

Chapter 3
A Follow-up Study of Reye's Syndrome in Czechoslovakia 43
I. Introduction 43
II. Reye's Syndrome Cases in the Years 1958 to 1970 43
  A. Clinical and Laboratory Data 43
  B. Morphological Features 46
    1. Gross Pathological Findings 46
    2. Histological Findings 46
  C. Bacteriological, Virological, and Toxicological Findings 47
  D. Epidemiological Data 54
III. A 15-Year Study of Reye's Syndrome in Relationship with Aflatoxin,
  1972 to 1986 54
  A. Material 54
  B. Method 55
  C. Chemical Analysis 55
    1. Thin Layer Chromatography (TLC) and Spectrophotometry 55
    2. Radioimmunoassay (RIA) 57
  D. Clinical and Laboratory Findings 57
  E. Epidemiological Data 59
  F. Aflatoxin in Reye's Syndrome Cases 60
  G. Biological Test 61
  H. Morphological Findings 61
IV. Reye's Syndrome with Chronic Liver Damage 66
V. Reye's Syndrome in the Newborns 84
VI. Experimental Models of Reye's Syndrome 94
  A. Aflatoxin B_1 and Influenza A Virus Interaction in Mice 94
    1. Materials and Methods 94
    2. Results 95
  B. Ultrastructural Changes in Interaction of Adenovirus 3
    and Aflatoxin B_1 on Tissue Cultures of HeLa Cells 97
    1. Materials and Methods 97
    2. Results 98
  C. Comments and Conclusion 107
References 110

Chapter 4
Carcinogenic Effect of Aflatoxin on Man 113
I. Introduction 113
II. Primary Liver Cancer in Third World Countries 113
  A. Factors Implicated in Liver Carcinogenesis 113

		1.	Aflatoxin 113
		2.	HBV Infection 114
		3.	Synergistic Mechanism Between Aflatoxin and HBV 115
		4.	Additional Risk Factor 116
III.	Primary Liver Cancer in the Western Countries 116		
	A.	Factors Implicated in Liver Carcinogenesis 116	
		1.	Cirrhosis 116
		2.	Alcohol 117
		3.	HBV Infection 117
IV.	Case Control Study of Liver Cancer in Czechoslovakia 117		
	A.	Materials and Methods 118	
		1.	HBsAg Determination 118
		2.	Aflatoxin Determination 118
	B.	Results and Comments 118	
		1.	Histological Investigation 120
		2.	HBV Infection 120
		3.	Alcohol Consumption 124
		4.	Aflatoxin in Liver Cancer Patients 124
		5.	Results of ELISA 127
		6.	Sources of Aflatoxin 128
	C.	Conclusion 129	
References 129			

Chapter 5
Potential Professional Risk Due to Aflatoxin Exposure Via the Respiratory Route 135

I.	Aflatoxin in Respirable Corn Dust Particles 135		
II.	Aflatoxin in Lung Tissue of Exposed Workers 136		
	A.	Case Reports 136	
	B.	Comment 146	
III.	Conclusion 148		
References 149			

Index 151

Chapter 1

MYCOTOXINS AND HUMAN DISEASES

I. HISTORY OF DISEASES DUE TO MYCOTOXINS IN MAN

The history of the role of mycotoxins in human diseases has been known since the ancient times, but scientific interest and systematic research in mycotoxins was not aroused until some 25 years ago.

Mycotoxins may be defined as toxic compounds produced by certain fungi growing on and contaminating feed and food. Fungi that produce mycotoxins are usually classified as molds. In defining the role of mycotoxins in medical science, a difference must be made between fungal metabolites elaborated under laboratory conditions — known as antibiotics — which are toxic to lower forms of life, such as bacteria, and those which occur naturally in the environment and are toxic to higher forms of life, including animals and even man himself. Whereas the discovery of antibiotics 40 years ago brought a remarkable advance in medicine, the naturally occurring mycotoxins present a great number of socioeconomic, agricultural, and veterinary problems.

The worldwide research on mycotoxins has made remarkable progress in recent years, but there still remain bewilderingly complex problems regarding their involvement in human health, although the health risk of certain fungi for man has been known for a long time.

Probably, the first mycotoxicosis to have been recognized in man was ergotism. This has its origin in the ingestion of rye and other grains contaminated with the mold *Claviceps purpurea*.[1] The first historical report of an epidemic of ergotism is known from Sparta in 450 BC.[2] During the Middle Ages, the disease was known in Europe as "St. Anthony's Fire" or the "Holy Fire", and periodic outbreaks of ergotism resulted in thousands of deaths. Epidemics of ergotism in this century are rare; however, serious outbreaks occurred in Russia (1926), England (1928), and France (as recently as 1951).[1]

Another mycotoxicosis recognized to have seriously affected both people and animals is Alimentary Toxic Aleukia (ATA), caused by metabolites from strains of the mold *Stachybotrys atra*. The incidence of this disease was recorded in Russia from time to time, beginning probably in the 19th century, but there are no published reports. When the "unknown horse disease" was being studied in the Ukraine, research workers headed by Drobotko[3] noticed that straw which had been fed to horses before an outbreak of the disease was overgrown with the fungus *S. alternans*. Experimental feeding of fodder infected with fungus to horses reproduced the pattern of the "unknown disease", which thereafter began to be known as "Stachybotriotoxicosis". Linnik[4] has established that workers handling musty straw suffered from some symptoms of this disease. The dominant pathological changes were necrotic lesions of the oral cavity, esophagus and stomach, and in particular, a pronounced leukopenia developing into a pancytopenia with dramatic hemorrhagic symptoms and death.

Although the disease received many names (septic angina, alimentary mycotoxicosis, aplastic anemia, hemorrhagic aleukia, and agranulocytosis), a committee of the Soviet Health Ministry concluded that the most appropriate term was ATA.

The most dramatic epidemics of ATA with a high mortality occurred in the Soviet Union in the war and postwar years of 1942 to 1947. The disorder developed after eating overwintered moldy grain, chiefly millet, but also wheat, rye, and oats. Studies on ATA from 1931 to 1943 have been reviewed by Sarkisov,[5] and more recently by Bilai[6] and Leonov.[7]

An association was established with the ingestion of grain contaminated by some molds, in particular *Fusarium poae* and *F. sporotrichoides*. Effects similar to ATA have been reproduced in cats fed on cultures of these species.[6] With improved harvesting and food production, the disease disappeared, and no other outbreaks have occurred.

The "yellow rice disease", described by the Japanese,[8] caused many human deaths. It was associated with the consumption of moldy rice imported from Southeast Asia. The disease involved acute cardiac beriberi accompanied by vomiting, ascending paralysis, convulsions, and respiratory arrest. Clinical features of the disease as well as epidemiological evidence and animal experiments with the toxin citreoviridin, produced by the mold *Penicillium citreoviride*, appeared to separate the cases of acute cardiac beriberi from those of pure vitamin B_1 deficiency. Improved techniques of harvesting and storage of rice prevented further outbreaks.

The toxic metabolite of the fungal genera *Aspergillus ochraceus* and *P. viridicatum* — ochratoxin A, identified in peanuts, corn, wheat, oats, barley, and rye — has been reported as the main cause of the mycotoxin porcine nephropathy, as seen in Scandinavia.[9]

The renal lesions included degeneration of the tubules, interstitial fibrosis and, at later stages, hyalinization of glomeruli, with impairment of the tubular function as a prime manifestation.[10] Attention has been called to the striking similarities in histological changes and functions of this animal disease to Balcan endemic nephropathy, which is a kidney disease observed in rural populations of Bulgaria, Rumania, and Yugoslavia.[11-13] Balcan endemic nephropathy is a chronic disease that is most common in people between 30 and 50 years of age and progresses slowly up to death. The disease occurs endemically, more often in females than in males.

The kidneys are remarkably reduced in size and histologically tubular degeneration, interstitial fibrosis, and hyalinization of glomeruli are present.[14] Impairment of the tubular function is a prominent and early sign of the disease.[11] Ochratoxin A has been found in blood samples of the people living in endemic nephropathy regions.[15] Epidemiological evidence of geographical distribution of the disease and a survey of foodstuffs suggest that ochratoxin A may be involved in Balcan nephropathy as the causal agent.[16]

II. AFLATOXINS

Despite the early recognition of ergot poisoning and toxicoses associated with yellow rice and overwintered grain, the mycotoxicoses remained a generally neglected category of diseases until 1960, when an unknown disease —Turkey X disease — caused the death of at least 100,000 turkeys in England.[17] A similar type of disease was also reported in ducklings, pigs, and cattle.[18,19]

The presence of Brazilian groundnut meal in animal rations was found to be the common factor in the disease outbreaks. The toxic factors were extracted from groundnut meal[20] and identified as metabolites of the common fungus *Aspergillus flavus*, which grew on the groundnut and gave rise to the name Aflatoxin, from *Aspergillus Flavus toxin*.

The extraction of a purified toxic substance from cultures of *A. flavus* was reported by two groups in 1962,[21,22] and the chemical structures were determined shortly thereafter by Asao et al.[23,24] and Chang et al.[25]

Although aflatoxin has been known for less than 15 years, its potential hazard to human health has caused great concern and has resulted in aflatoxin becoming one of the most studied mycotoxins in the human environment. A large number of researchers and vast resources were drawn into investigations of the problem, and as a result, an enormous amount of literature in the field has accumulated during the past 20 years.

In early studies on the isolation of the toxin, Nesbitt[22] obtained two types: aflatoxin B and aflatoxin G. The former emits blue fluorescence (aflatoxin B), whereas the latter emits green fluorescence (aflatoxin G). Hartley and colleagues[26] showed that aflatoxin could be separated into four closely related compounds: aflatoxins B_1, B_2, G_1, and G_2. More recently, related substances initially isolated from the milk of cows fed aflatoxin-contaminated feed were designated as aflatoxin M (milk toxin).[27,28] Although a total of 12 structurally related compounds with similar configurations have been identified and all of them designated as aflatoxins, the term aflatoxins usually refers to four compounds of the group of bisfurano-coumarin metabolites:

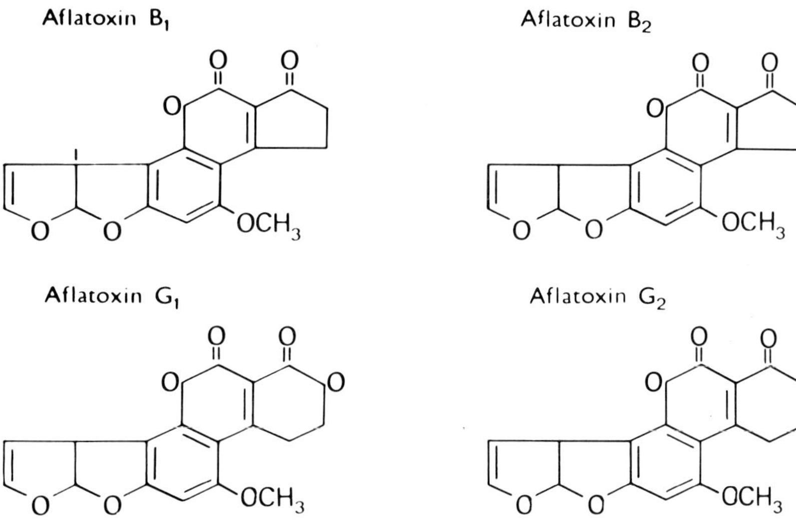

FIGURE 1. Chemical structure of aflatoxins.

B_1, B_2, G_1, and G_2. (Figure 1 demonstrates the chemical structure of aflatoxins B_1, B_2, G_1, and G_2.)

The relative proportions of the four major aflatoxins produced by *Aspergillus* cultures vary in both the genetic constitution of the fungus and in the environmental parameters associated with fungal growth. In general, strains of *A. parasiticus* are highly toxigenic and regularly produce B_1, B_2, and G_2, while *A. flavus* strains produce low to high levels aflatoxins B only.[29] Aflatoxins are freely soluble in moderately polar solvents (e.g., chloroform and methanol) and especially in dimethylsulfoxide (DMSO), which is used as a vehicle in the administration of aflatoxins to experimental animals. As a pure substance, the aflatoxins are very stable at high temperatures. They are relatively unstable when exposed to light and ultraviolet (UV) radiation. Little or no destruction of aflatoxins occurs under ordinary cooking conditions and during pasteurization. However, roasting groundnuts reduces the level of aflatoxins.[30]

A. OCCURRENCE AND SOURCES OF AFLATOXINS

The molds *A. flavus* and *A. parasiticus* are ubiquitous all over the world except in polar regions. Their growth and production of toxins require humidity and temperature conditions that are prevalent in tropical and subtropical areas, but may be occasionally found in colder regions, such as Europe and the U.S.A. The formation of aflatoxins mainly takes place during harvesting and storage, although field crops may be invaded by *A. flavus* before the harvest by insects carrying fungal spores, which can result in a preharvest formation of toxins.[31] Aflatoxin-producing molds grow on a large variety of foodstuffs, particularly on plant products, but it appears that certain foodstuffs are more suitable substrates than others. These include oil seeds (groundnuts, peanuts, Brazil nuts, pistachio nuts, cottonseed, almonds, walnuts, and copra)[32,33] and some cereals (maize, rice, wheat, barley, oats, and sorghum).[32,34] Aflatoxin has also been found in Indian red peppers[35] and fig fruits.[36]

Surveys in several countries have shown that aflatoxin M_1 may be present in liquid or dried milk and milk products.[37-39] It has been shown that the level of AFM_1 in milk is directly related to the daily intake of AFB_1 in dairy feeds.

Residues of AFB_1 have also been found in the liver and musculature of poultry[40] and in the liver, kidney, and musculature of swine.[41-43] AFB_1 has been demonstrated in eggs and tissue of laying hens fed aflatoxin.[44,45] It can be seen that the range of foodstuffs which may be contaminated by aflatoxins is very large. On the one hand, in vegetable foods, this contamination results directly from fungal spoilage; on the other hand, milk, meat, and eggs become contaminated indirectly through the absorption of aflatoxin-contaminated feed by farm animals, which results in residues of the parent toxin or its metabolites in body tissues or fluids.

B. METHODS OF ANALYSIS

Two types of tests have been developed for the detection and determination of aflatoxin: biological and chemical.

Biological tests — Since its discovery, aflatoxin has been evaluated in a great variety of biological systems, including subcellular enzyme systems, cell cultures, microorganisms, and laboratory animals. In the original biological test,[46] 1-d-old ducklings were used as test animals for determining the presence of aflatoxin in suspect food with bile duct proliferation as a semiquantitative index. Burmeister and Hesseltine[47] examined over 300 species of microorganisms for their sensitivity to aflatoxin and found that *Bacillus brevis* and *B. megaterium* were most sensitive to aflatoxin. Of all biological tests, the commonly used method is the chicken embryo bioassay, in which 0.1 to 0.2 µg of AFB_1 is applied to the egg and the mortality rate recorded during the 23-d period of hatching.[48]

Chemical methods — Numerous methods of analysis have been reported for the determination of aflatoxin in human and animal foodstuffs and have been described in AOAC's "Official Methods of Analysis".[49] Existing methods for determining aflatoxins are based on an extraction with aqueous acetone or methanol, extract clean-up by column chromatography, then quantitation after thin-layer chromatography (TLC), generally two-dimensional. These methods are often limited by low precision and the relatively low chromatographic efficiency of the TLC methods. High-performance liquid chromatography (HPLC) has now been widely applied to the separation and determination of aflatoxins because of its high degree of sensitivity, precision, and accuracy.[50,51]

Recent investigations have described the development of specific antibodies for aflatoxins and the efficacy of their use in radioimmunoassay (RIA)[52,53] and enzyme-linked immunosorbent microassay (ELISA).[54,55] These immunological methods offer the possibility of detecting aflatoxins at very low amounts and may be used for routine screening and monitoring of food as well as of human exposure.[55,56]

C. TOXICITY

The toxicity of aflatoxin in different animal species was reviewed by Allcroft,[57] Newberne,[58] and Butler.[59] The toxic effect of aflatoxin has been demonstrated in many animal species. Rainbow trout, ducklings, turkeys, guinea pigs, rabbits, and dogs are highly susceptible, while sheep were found to be the most resistant among the animal species. It was also reported that AFB_1 can cause a significant liver injury in primates.[60] Although interspecific variation has been detected for acute effects, many factors, such as the age, sex, nutritional status, and mode of application, affected the degree of toxicoses. In general, aflatoxin is more toxic for young animals and males than females.

In all the animals studied, the liver was the principal target organ. The first signs of aflatoxicosis in animals were the lack of appetite and the loss of weight. Liver centrilobular necrosis, fatty degeneration, and bile duct proliferation were the most common pathological findings. Besides the liver, many other organs were more or less severely affected in acute experiments with high doses of aflatoxins.[61] Congested lungs and occasional necroses in the myocardium and kidneys were observed during the first days following administration. With the exception of the liver damage, in which fibrosis with bile duct proliferation was present, all the other organs appeared normal in surviving animals. For most of the animals tested, the LD_{50} for a single dose of AFB_1 was in the range of 0.5 to 10 mg/kg of body weight.[58]

Aflatoxin has also been reported to be toxic to tissue cultured cells. The LD_{50} of AFB_1 for chicken embryo primary cultured cells and HeLa cells is 5 ug/ml,[62] and the concentration of 0.05 ug/ml inhibits the growth of human embryonic lung cells.[63] AFB_1 was also found to be toxic to cultured adult human liver cells.[64]

D. CARCINOGENICITY

The carcinogenicity of aflatoxin was first reported in 1961, when Lancaster et al.[65] showed that rats developed hepatomas when fed the same peanut meal implicated in Turkey X disease. Subsequently, in tests with several laboratory species, including nonhuman primates, AFB_1 has been identified as the most potent hepatocarcinogen known.

A correlation between dietary aflatoxin content and liver tumor incidence was reported by Wogan et al.[66] in male Fischer rats fed dietary concentrations of AFB_1 ranging from 1 to 100 µg/kg. A 10% tumor incidence was found after 104 weeks at 1 µg/kg, compared with 100% tumor incidence in a group fed AFB_1 at a dose of 100 µg/kg after 54 weeks.

AFB_1 is hepatocarcinogenic to rainbow trout at very low dietary levels. A diet containing 20 ppb AFB_1 administered for 9 months induced liver tumors in more than 50% of the fish.[67] Studies on the carcinogenic activity of AFB_1 in ducks showed that 8 out of 11 animals had developed hepatic tumors after being fed for 14 months on a diet containing 30 ppb aflatoxin from toxic groundnut meal.[68] Carcinogenic effects of AFB_1 have been reported also in nonhuman primates. A female rhesus monkey developed primary liver cancer after ingesting a total of about 500 mg of AFB_1 over a 6-year period.[69] Of the two rhesus monkeys treated with AFB_1 for over 5 years at the National Institute of Nutrition, Hyderabad, India, one (a male) developed hepatocellular carcinoma,[70] while the other (a female) developed a cholangiocarcinoma.[71] Reddy[72] reported that intermittent feeding on a diet containing 200 µg AFB_1 per kilogram of the diet produced hepatocellular carcinomas in 6 out of 6 females (100%) and in 3 out of 6 males (50%) of 12 tree shrews *(Tupaia glis)* after 74 to 172 weeks of treatment. In two tree shrews, the liver tumors were associated with extensive postnecrotic scarring of the remaining hepatic parenchyma, whereas in the other seven tumor-bearing livers, only mild to moderate stellate fibrosis of portal tracts was encountered.

Although in laboratory animals the liver is a primary target organ for AFB_1, tumors have been demonstrated at sites other than liver. Sieber et al.[73] observed various neoplasms in Old World monkeys (rhesus, cynomolgus, and African green) treated with an average total AFB_1 dose of 709 mg for an average of 114 months. In 13 out of 35 autopsied animals, only five liver tumors (two hepatocellular carcinomas and three hemangioendothelial sarcomas) were found. Two animals had osteogenic sarcomas, six had carcinomas of the gall bladder, three had tumors of the pancreas, and one had papillary carcinoma of the urinary bladder. Carcinomas of the stomach[74] and mucinous adenocarcinomas of the colon[75] were also reported. A high incidence of renal epithelial neoplasms was observed in male Wistar strain rats fed on diets containing AFB_1 levels of 1.0, 0.5, and 0.25 mg/kg feed for 147 d.[76] Other extrahepatic tumors were reported, such as squamous cell carcinoma of the tongue[77] and esophagus.[78]

Aflatoxin G_1 has also been found to be a very potent carcinogen. It was reported that feeding AFG_1 to rats at a dose of 3 µg/ml in drinking water resulted in liver tumors in 21 out of 26 rats after the total dose of 6 mg. Of these 26 animals, 6 also developed kidney tumors.[78]

Aflatoxin M_1 induces liver cell carcinomas in rainbow trout[79,80] and in rats,[81] but is considerably less potent than B_1 in both species.

Aflatoxin B_2 is weakly active in inducing liver tumors in rats; doses more than 100 times higher than an effective dose of AFB_1 are required.[82]

E. TERATOGENIC ACTIVITY

Butler and Wigglesworth[83] studied the effects of AFB_1 on pregnant rats and found that oral ingestion of the toxin caused fetal growth retardation. Animals dosed in early pregnancy showed only a slight reduction in placental weight. Those dosed on day 16 after fertilization showed severe growth retardation. The authors concluded that weight reduction was a secondary result of aflatoxicosis, and the hepatic damage in mothers caused by aflatoxin was thought to be the cause of the reduction of fetal weight. Butler[84] in a follow-up study showed that the reduction in growth was only indirectly related to the severity of the hepatic damage and concluded that the reduced fetal weight were a result of reduced food consumption by the mothers.

Elis and DiPaolo[85] reported that AFB_1 is a potent teratogen for hamsters. A single intraperitoneal injection of AFB_1 at 4 mg/kg body weight, given on day 8 of pregnancy, resulted in a high proportion of malformed and dead or reabsorbed fetuses. As to Butler's[84] findings, there was severe maternal hepatic damage. The malformations observed in early treatments were anencephaly, disorganized cranial end of the neural tube, severe growth retardation, and ectopia cordis. Later treatments showed microcephaly and umbilical hernia. In rats and mice, DiPaolo et al.[86] reported that days 9 to 11 of gestation were the most susceptible to the action of teratogenic compounds.

Other studies on chicken embryos,[87-89] showed that AFB_1 and several metabolic intermediates were capable of producing various teratogenic effects on the developing fetus. Edema and growth retardation were evident after an injection of the mycotoxin into the air sac. However, Hintz et al.,[90] who studied the teratogenic effects of aflatoxin on sows given feed containing 450 ppb of AFB_1, did not find malformations in the piglets. Therefore, it has been suggested that AFB_1 is capable of teratogenic effects only in some experimental species. Differences in metabolic pathways of aflatoxin between different animals may contribute to such specificity.

F. MUTAGENICITY

AFB_1 is known to be a potent mutagen. The first study on the induction of chromosomal aberrations by aflatoxin was reported by Lilly.[91] In his experiment on the roots of *Vicia faba* seedlings, treated at 21°C for 3 h with a 67 mM mixture of aflatoxins, it was found that aflatoxin causes a significant increase in the number of abnormal anaphases. The abnormalities consisted of chromosomal fragments with occasional bridges. Cultured human leukocytes with a similar aflatoxin mixture produced a high frequency of chromosomal aberrations. Chromatid breakage was the most common aberration, while no chromosomal breakage was found.[92]

Other experiments showed that the percentage of cells with chromosomal aberrations depended on the level of the toxin concentrations as well as on the duration of the treatment.[93] AFB_1 was shown to induce mutations in transforming DNA of *Bacillus subtilis*,[94] and was mutagenic to vegetative cells (but not conidia) of *Neurospora crassa*[95] and to *Chlamydomonas reinhardii*.[96] Aflatoxin was also shown to induce recessive lethal mutations in *Drosophila melanogaster*.[97] When applied directly to test strains of *Salmonella typhimurium*, aflatoxin showed no biological effect,[98] but an incubation of AFB_1 with a rat or human liver microsomal system in the presence of test strains of *S. typhimurium* produced reverse mutations in the bacteria.[99-101] The studies indicated that a metabolic activation of aflatoxin was necessary for the expression of biological activities.[99-102]

Toxicity of the aflatoxin metabolite to bacteria decreased when DNA or RNA was added to the assay system. The binding was found to be more effective when DNA was used instead of RNA.[101] Several studies have shown that the reactive metabolite of aflatoxin B_1 binds covalently to the nucleic acids. Later studies of Garner[103] and Swenson et al.[104] indicated that AFB_1 is converted by the hamster or rat liver *in vivo* and by human liver microsomes *in vitro* to a highly reactive compound that is probably the ultimate carcinogen, aflatoxin B_1-2,3-oxide.

G. IMMUNOLOGICAL EFFECTS

One of the pronounced effects of aflatoxin is its ability to impair the immune system. Aflatoxin was reported to increase the severity of avian infectious diseases such as salmonellosis,[105] crop mycosis,[106] aspergillosis,[107] cecal coccidiosis,[108] and Marek's disease.[109]

In chickens, aflatoxin caused a dose-related suppression of the primary humoral response of antibodies and regression of the thymus and bursa Fabricii, which are primary determinants of immunocompetence.[110] Richard and Thurston[111] showed that aflatoxin reduced phagocytosis of *Aspergillus fumigatus* spores by alveolar rabbit macrophages, and Michael et al.[112] reported that it reduces the effectiveness of the chicken reticuloendothelial system.

Aflatoxin B_1 added to cultures of peritoneal rat macrophages and mouse fibroblasts inhibited the uptake and incorporation of tritiated leucine and uridine by both cell types and impaired the phagocytic activity of macrophages.[113] Impaired phagocytosis by chicken heterophils during aflatoxicosis was reported by Chang and Hamilton,[114] who found impaired spontaneous and chemotactic locomotion and reduced ability of heterophils to kill bacteria. Aflatoxin was also reported to decrease significantly the complement activity.[112,115] The effect of aflatoxins (B_1, B_2, G_1, G_2) on interferon induction by the influenza virus in cell monolayers was studied by Hahon et al.[116] Out of the four aflatoxins, AFB_1 was the most deleterious to both cell growth and the viability of cells in confluent cultures. When the influenza virus was assayed by using cells previously treated with aflatoxins, virus assay values were comparable to those of controls. In subsequent experiments, however, virus growth rates in the aflatoxin-treated and normal cells were comparable, but the attained levels of virus concetration were two- to fourfold higher in the cell cultures previously treated with aflatoxin. The magnitude of virus growth was related to the concentrations of aflatoxin. The authors suggested that aflatoxin may exert its adverse effect on interferon induction through (1) formation of the inducing molecule, (2) activation or depression of the interferon gene, or (3) transcription and translation of the interferon messenger RNA.

H. METABOLIC ACTIVATION OF AFLATOXIN

Aflatoxins are primarily metabolized by microsomal mixed-function oxidases, resulting in a variety of detoxification and activation products. Because these enzymes are found in the highest concentrations in the liver, it is not surprising that the liver is a target organ for the toxic and carcinogenic effects of aflatoxins.

Aflatoxin M_1 (AFM_1) was the first identified biotransformation metabolite deriving its name from its presence in milk. The original observation by Allcroft and Carnaghan[27] was that milk from aflatoxin-fed cows contained a toxic substance capable of producing liver damage in 1-d-old ducklings, typical of AFB_1. Butler and Clifford[117] then found that this toxic substance was present in the liver of rats given pure AFB_1, indicating that some of the AFB_1 was converted in the liver. Allcroft et al.,[28] after isolating the "milk" toxin from the urine of sheep fed AFB_1, proposed the name aflatoxin "M".

The toxicity of AFM_1 appears to be equivalent to that of AFB_1,[118] although the carcinogenicity[119] and mutagenicity[120] were considerably smaller than those observed in AFB_1.

1. Aflatoxin B_2a and G_2a

Liver homogenates of certain avian and rodent species were shown to convert AFB_1 and G_1 to their 2-hydroxy,2,3-dihydroderivatives called aflatoxins B_2a and G_2a.[121] These metabolites may represent major metabolites which strongly bind to protein and are probably sufficiently reactive when formed *in vivo* to cause acute toxic effects.[122]

2. Aflatoxicol

Aflatoxicol[123] occupies a special position for it is not formed in a microsomal mixed-function oxidase reaction, but by an NADP-linked dehydrogenase of the cytosol, which also has a 17-ketosteriod dehydrogenase activity. It is the only metabolite transformation of aflatoxin *in vitro* known to be sensitive to hormones.[124] Furthermore, this reaction is reversible and may provide a "reservoir" for AFB_1 and its metabolites.[125] Its toxicity, measured by lethality for *S. typhimurium* in the presence of mitochondrial fractions of rats, has indicated that its potency is lower than that of AFB_1, but higher than AFM_1 potency.

3. Aflatoxin P_1

Aflatoxin P_1 was found in a major urinary metabolite in rhesus monkeys given a single intraperitoneal injection of ring-labeled AFB_1.[126,127] AFP_1 is produced by the NADPH-dependent mixed-function oxidase enzyme system of liver microsomes. It is not toxic in the chick embryo test system,[89] nor is it lethal in the *S. typhimurium* test system.[100]

4. Aflatoxin Q_1

Aflatoxin Q_1 represents approximately $1/_3$ of the metabolites produced from AFB_1 by microsomes of the monkey and human liver.[128,129] Hsieh et al.[130] observed that AFQ_1 was 18 times less toxic than AFB_1 and was not mutagenic in bacterial mutagenesis tests using *S. typhimurium*.

5. Highly Active Forms of AFB_1

Metabolic activation was shown to be required for covalent binding of AFB_1 to DNA, suggesting that the formation of the 2,3-epoxide intermediate might be involved in its biological activity,[131] and that this covalently bound aflatoxin may be the basis for its toxicity and carcinogenicity.[132] The variation in species response to the carcinogenic action of AFB_1 was correlated with covalent binding to nucleic acids,[133] metabolism,[134] and mutagenicity of metabolites.[102] In the initial report, Garner and co-workers[98,100] proposed that the livers of most, if not all, animal species are capable of converting AFB_1 to unidentified toxic metabolite(s), which could be assayed in a bacterial mutagenicity test. It was suggested that this metabolic product might be the 2,3-oxide existing transiently and alkylating in bacterial DNA. Indirect evidence for this epoxidation as a critical metabolic reaction was obtained by Swenson[104] on the isolation of 2,3-dihydro-2,3-dihydroxy-AFB_1 after acid hydrolysis of the nucleic acids extracted from the livers of rats injected with AFB_1 and from the nucleic acids treated with AFB_1 in the presence of rat and human microsomes. This has been assumed to be strong indirect evidence of the formation of the 2,3-oxide, and, in view of the interaction with DNA, it is now generally accepted that the epoxide of AFB_1 is the bacterial mutagen and the proximal carcinogen.

From among the above-mentioned aflatoxin-induced biotransformation products, AFB_1-epoxide and AFB_2a are considered the active forms of AFB_1, whereas the others are considered detoxification products. However, AFM_1 and aflatoxicol are still quite toxic and carcinogenic.[135]

I. BIOCHEMICAL EFFECTS OF AFLATOXIN

Aflatoxin is metabolized in the endoplasmic reticulum. Interaction between activated molecular species and the liver cell apparently occurs at several loci. In the nucleus, DNA-dependent RNA polymerase is inhibited,[136] the toxin binds covalently to DNA *in vitro* and *in vivo*,[101,137] DNA repair is stimulated,[138] and the aflatoxin is activated on the outer nuclear membrane to a form that inhibits RNA synthesis.[139] The permeability of mitochondria is increased and electron transport is interrupted with a decline in respiration.[140] Alterations of nuclear RNA metabolism resulting from AFB_1 are associated with changes in the morphologic characteristics of the nuclei as observed in electron microscopy.[136]

Responses of other metabolic and biochemical parameters to aflatoxins have also been reported. In lipid metabolism, Wei et al.[141] observed that acute and chronic administrations of AFB_1 to rats caused a significant suppression of acetate incorporation into the total lipids of hepatic and adipose tissues. Kato et al.[142] achieved results analogous with acetate incorporation into hepatic cholesterol in rats. Williams and Rabin[143] first established that incubation *in vitro* of AFB_1 and of hepatic microsomes of rats resulted in a displacement of polysomes, and that microsomal membranes lost their ability to bind polysomes. This effect was antagonized by corticosterones. The effects of aflatoxins on lysosomes were studied by Pokrovsky et al.,[144] who reported increases in five lysosomal enzymes and suggested that their effects on cellular structure may be a component of the toxic mechanism of aflatoxin. Similar results were reported by Tung et al.,[145] who found that feeding aflatoxin-contaminated diets caused increases in acid phosphatase in the liver of chickens. Altered glucose metabolism was also reported.[146]

J. FACTORS MODIFYING THE EFFECTS OF AFLATOXINS

Numerous reports have dealt with various factors that modify the carcinogenic and toxic effects of aflatoxins in experimental animals. These include:

1. Sex-linked differences
2. Endocrine status
3. Interaction with other environmental factors, particularly the nutritional factors

1. Sex Differences

Several studies have indicated that in comparison with males, females are more resistant to both acute toxic and carcinogenic effects of aflatoxins. Newberne and Wogan[147] observed a considerably longer period between the appearance of hepatic precancerous lesions and progression to liver carcinoma in female Fischer rats compared to males fed diets containing the same level AFB_1. Butler[61] estimated LD_{50} for AFB_1 to be 7.2 mg/kg of body weight for male and 17.9 mg/kg of body weight for female rats. In a later study by Ward et al.,[77] male rats kept on a diet containing AFB_1 at 2 mg/kg died of malignant hemorrhagic liver tumors significantly earlier than females.

2. Endocrine Status

An important sex hormone effect on aflatoxin carcinogenesis was demonstrated by Newberne and Williams[148] in male rats fed for more than 1 year on a diet containing AFB_1 at 0.2 mg/kg and diethylstilbestrol at 4 mg/kg of body weight. Liver tumors developed only in 8/40 of these animals, whereas liver cancer appeared in 25/35 of the control group fed the same diet without estrogen. The protective effect on aflatoxin carcinogenesis has been demonstrated in hypophysectomized rats. In a group of 14 hypophysectomized rats, none developed liver tumors when fed a diet containing an AFB_1 level of 4 mg/kg of body weight, whereas 11/14 control rats developed liver tumors in 49 weeks.[149] An inhibitory effect on the aflatoxin activity has been reported in rats given adrenocorticotropin (ACTH) 4 U weekly and AFB_1 in the amount of 125 µg/animal weekly for 20 weeks. None of these animals developed liver cancer in contrast to 100% of liver tumors in rats receiving AFB_1 alone. It has been suggested that the protective action of ACTH toward AFB_1 hepatocarcinogenesis may be related to adrenal stimulation, because corticosteroids appear to affect the binding of polysomes to the endoplasmic reticulum membranes.[150] The evidence that AFB_1 and steroid hormones are competitors for the same binding sites on the endoplasmic reticulum and that they act in an antagonistic manner was originally described by Williams and Rabin.[143]

3. Nutritional Factors

There are numerous conflicting reports on the effect of dietary protein, deficiency in lipotrope, vitamin B_{12}, and choline on the toxicity and carcinogenicity of aflatoxins.[151-155] Madhaven and Gopalan[151] found that high dietary protein could enhance aflatoxin-induced hepatoma in the rat liver. Wogan and Newberne[75] reported a high incidence of liver cancer in rats continuously fed a synthetic diet high in protein and vitamin B_{12} to which aflatoxins were added. In 1969, Rogers and Newberne[154] observed that the number of hepatic tumors found in lipotrope-deficient adult rats fed a carcinogenic dose of AFB_1 was lower than the number in a control group. However, in 1971 they repeated their study using weanling rats and found that the incidence of tumors was higher in the lipotrope-deficient rats.[155,156] Temcharoen et al.[157] reported a high incidence of aflatoxin-induced hepatoma in the high-protein group of rats compared to the low-protein group in which a high incidence of hyperplastic nodules and cholangiofibrosis had been observed. Since these changes were presumably precancerous lesions, the authors suggested that they might have progressed to malignancy if the experiment had been extended, and that the ultimate result of aflatoxin-induced neoplasm in the low- and high-protein groups might not have differed.

Vitamin B_{12} was shown to enhance significantly the induction of both hyperplastic nodules and hepatoma by aflatoxin in the higt-protein group of rats, but the influence of dietary vitamin B_{12} on aflatoxin carcinogenesis in the low-protein group seemed negligible in the study of Temcharoen et al.[157]

Vitamin A deficiency has been found to enhance the mortality and acute liver damage in aflatoxin-treated male rats in contrast to minimal liver damage in female rats and male rats given vitamin A.[158] In a study conducted by Newberne and Rogers,[159] it was shown that rats fed diets containing AFB_1, deficient, adequate, or excessive in vitamin A, developed a similar incidence of liver tumors, but the vitamin A deficient group had an increased incidence of colon carcinomas. However, excessive vitamin A did not protect the animals against carcinogenesis in either the liver or the colon.

Dietary fat influences the response of the rat liver to aflatoxin.[160] Corn oil, compared to beef fat, resulted in a significant enhancement of liver tumors. The increased incidence of liver tumors associated with corn oil coincided with an increased activity of the liver microsomal enzyme.

Several studies have demonstrated that certain dietary metals provide a protective effect on aflatoxin toxicity in several animal species. Newberne and Conner[161] first reported that selenium supplementation up to a dietary level of 1.0 ppm reduced the acute toxicity of AFB_1 in rats, while greater levels enhanced their mortality. The protective effect of selenium was confirmed in several studies on pigs and turkeys.[162,163] Aleksandrowicz et al.[164] demonstrated that sodium selenate added to the culture medium with AFB_1 inhibited blastoid transformation of human lymphocytes. A protective effect on AFB_1 activity was also reported when cadmium was added to a diet for young pigs.[165]

III. EFFECTS OF AFLATOXIN ON MAN

A. EXPOSURE CONDITIONS

Association of an outbreak of liver damage in turkeys with aflatoxin was of basic importance in the recognition of aflatoxin as an environmental hazard. The following discovery that aflatoxin was toxic and carcinogenic in a broad range of animal species led to the speculation that it also might be hazardous for man.

The main source of human exposure to aflatoxin is presented by contaminated food. In tropical areas where the staple food is often heavily contaminated with this mycotoxin and the population is exposed to monotonous diets based on such staple foods, the health risk is particularly high. However, aflatoxin has been occasionally found in various food substrates and food products in Europe and the U.S., indicating that the problem of aflatoxin is not limited to any one geographical area, but is a real or potential problem all over the world.

Although humans and animals are primarily exposed to aflatoxin via the diet, considerable evidence underlies the possibility that aflatoxin present in respirable particles may pose a potential hazard via the respiratory route, particularly in workers in agricultural settings.[166-169]

A recent study showed that dermal exposure to aflatoxin could also pose a potential risk, particularly to workers in research laboratories.[170]

B. ACUTE EFFECTS OF AFLATOXIN

Acute aflatoxicosis in man has rarely been reported. Probably the first sporadic cases of acute aflatoxicosis in humans were reported by Ling et al.[171] in 1967. They observed three children in the Province of Taiwan who died of acute liver necrosis associated with the ingestion of rice contaminated with aflatoxin at a level of 200 µg/kg. One year later, Van Walbeck et al.[172] described an acute liver disease in members of one family in Canada associated with the ingestion of spaghetti heavily contaminated with aflatoxin. Serck-Hansen[173] reported fatal hepatitis in a 15-year-old boy in Uganda, associated with the ingestion of aflatoxin-contaminated cassava. Acute liver dystrophy in a man of 45 years, associated with aflatoxin-contaminated nuts, was reported by Bosenberg[174] in West Germany.

The first epidemic of aflatoxicosis affecting men and dogs has been described by Krishnamachari and co-workers[175] in India. The outbreak started in 1974 almost simultaneously in more than 150 villages of the states of Gujarat and Rajasthan, in northwestern India. In this epidemic, several hundred men exhibited signs and symptoms of a poisoning characterized by icterus, a rapid onset of signs of portal hypertension, and ascites. More than 100 people died. The infants were completely spared and children under the age of five were less commonly affected than adults. This outbreak was traced back to the consumption of maize contaminated with aflatoxin at levels up to 15 mg/kg. The examination of tissues and body fluids for aflatoxins was limited to 15 samples (one necropsy liver sample, seven urine samples, and seven blood samples). Aflatoxin was detected only in two blood samples.

Later a different group headed by Tandon[176] reinvestigated the outbreak in Rajasthan and reported 994 affected persons with 97 deaths. In another part of the study by Tandon et al.,[177] clinical data on 200 patients were analyzed. This included 176 patients admitted to a hospital and 24 patients still suffering from the disease. Approximately 81% of 176 hospitalized patients showed almost complete recovery within 2 to 8 weeks. Within 6 weeks after the onset, 10% of the patients died. The disease affected all age groups including children and both sexes. A peculiar and very notable feature of the outbreak was that the appearance of the disease in the village population was preceded by an occurrence of a similar disease in domestic dogs which almost resulted in fatality.

The significant changes in the liver obtained at the biopsy from eight patients and at the autopsies from one human and two dogs were almost identical, characterized by centrilobular scarring, severe bile duct proliferation, cholestasis, and a syncytial giant-cell transformation of the hepatocytes.

Detailed epidemiological studies indicated a high familial incidence. The staple food of the people, all agriculturists, was corn. They had consumed bread made from flour prepared from corn that had become moldy during damp storage for a few weeks preceding the onset of the illness. Samples of the moldy corn collected from the affected households yielded *Aspergillus flavus*. Aflatoxin detected in these samples was mostly AFB_1 and G_1.

Recently, an outbreak of acute hepatitis in 20 patients aged $2^1/_2$ to 45 years, caused by aflatoxin, has been reported by Ngindu and co-workers[178] in Kenya. Twelve of them died. The illness tended to occur in family groups. Two families, from which 8 of 12 sick members died, were eating maize which contained as much as 12,000 ppb of AFB_1. In the liver samples obtained from two patients at the autopsy, AFB_1 at concentrations of 39 and 89 ppb has been proved.

Lately, a relationship between kwashiorkor and aflatoxin was pointed out by Hendrickse and collaborators.[179,180] The authors investigated blood and urine samples from 252 Sudanese children for the presence of aflatoxins. The children comprised 44 with kwashiorkor, 32 with marasmic kwashiorkor, 70 with marasmus, and 106 age-matched, normally nourished controls. Aflatoxins were detected more often and at higher concentrations in sera from the children with kwashiorkor than in the other malnourished and control groups. In a later study,[180] clinical and field research in the Sudan was augmented by a study of autopsy liver specimens obtained in Nigeria, South Africa, and Liberia from children who had died of kwashiorkor and other forms of protein-energy malnutrition. A total of 469 sera and 468 urines were analyzed for aflatoxins. Aflatoxicol, a metabolite of AFB_1, was detected in kwashiorkor (12%) and marasmic kwashiorkor (6%), but not in the controls and only once in marasmus.

An epidemiological study and analysis of foods from local markets in Khartoum as well as in homes revealed widespread aflatoxin contamination. The authors conclude that the results of their studies establish the relationship between aflatoxin and kwashiorkor, the nature of which is still obscure, but included a possibility of causal association. This suggestion seems to be supported by a recent report of Long,[181] who observed a kwashiorkor-like disease in guinea pigs accidentally fed aflatoxin-contaminated feed.

Chronic changes following acute damage of the liver due to aflatoxin were described in children suffering from varying degrees of protein-caloric malnutrition who had accidentally consumed aflatoxin-contaminated commercially-produced peanut flour for periods ranging from 5 d to 4 weeks.[182] A histological examination of the liver biopsy showed a gradual transition from fatty metamorphosis in the first 2 months after ingestion.

Payet et al.[183] followed two Senegalese infants who received a peanut meal supplement for 10 months. Samples were later found to be contaminated with aflatoxin at 0.5 to 1.0 mg/kg. Both infants had chronic hepatitis, gross abnormalities persisting in one through his 6th year.

A relationship between the juvenile cirrhosis in India and aflatoxin was suggested by Robinson,[184] based on a blue fluorescent B_1 spot in the breast milk of mothers and the urine of children with this disease, but later studies have not provided sufficient evidence for this speculation.

C. CARCINOGENIC EFFECT OF AFLATOXIN

Epidemiological studies have consisted of estimates of aflatoxin intake by the population in which the incidence or prevalence of primary liver cancer (PLC) was determined simultaneously. In one such study in Uganda, the frequency of aflatoxin contamination of market food samples was positively associated with a liver cancer incidence in localized population groups.[185] A study in Thailand showed variations in aflatoxin intake over three areas of the country and a threefold difference in PLC incidence rates between two of them.[186] A similar study carried out in the Murang, a district of Kenya,[187] and in Swaziland[188] revealed a positive correlation between aflatoxin intake and liver cancer incidence. High levels of aflatoxin in food were shown to be associated with a high incidence of liver cancer in Mozambique.[189] The results of individual studies, particularly of those made in Kenya and Swaziland demonstrating a dose-response relationship, are indicated in Table 1.

Marked geographical differences in the incidence of primary liver cancer have recently been demonstrated in China.[191] Areas with the highest incidence of liver cancer are rural areas having higher rates of exposure to aflatoxin.

Although aflatoxin has particularly been recognized as a potent hepatocarcinogen, a wide variety of tumors at sites other than the liver were reported in experimental animals.[73,74,78] A possible relationship between aflatoxin and carcinoma of organs other than the liver in men has also been suggested. Deger[192] described two British biochemists who had developed adenocarcinomas of the colon after exposure to purified aflatoxins. One biochemist had worked with this material from 1962 to 1964 and developed symptoms in 1971. The second biochemist had done this work for 12 months between 1969 and 1970 and developed symptoms in 1972.

Two chemical engineers who had worked on a method for sterilizing peanut meal contaminate with *A. flavus* were reported to have died from alveolar carcinoma.[166]

Prospective studies of exposed workers are planned for determining the associated risk of a specific form of cancer and other diseases in man, particularly in relation to exposure primarily via the respiratory route. A recent report[169] of an excess of cancer deaths among workers at a Dutch oil-pressing facility, in which exposure to aflatoxin occurred primarily through inhalation, produced evidence for the potential professional health risk of aflatoxin.

TABLE 1
Summary of Available Data on Aflatoxin Ingestion Levels and Primary Liver Cancer Incidence[a]

Country	Area	Aflatoxin (estimated average daily intake in adults — ng/kg, body weight per day)	Liver cancer (no. of cases registered)	Incidence per 10^5 of total population per year
Kenya	High altitude	3.5	4	1.2
Thailand	Songkhla	5.0	2	2.0
Swaziland	High veld	5.1	11	2.2
Kenya	Middle altitude	5.9	33	2.5
Swaziland	Mid veld	8.9	29	3.8
Kenya	Low altitude	10.0	49	4.0
Swaziland	Lebombo	15.4	4	4.3
Thailand	Rarburi	45.0	6	6.0
Swaziland	Low veld	43.1	42	9.2
Mozambique	Inhambane	222.1		13.0

[a] From Peers and Linsell.[190]

REFERENCES

1. **Wogan, G. N.,** Alimentary mycotoxicoses, in *Food-Borne Infections and Intoxicants,* Riemann, H., Ed., Academia, New York, 1969, 395.
2. **Linsell, C. A.,** The mycotoxins and human health hazards, in *Mycotoxins in Foodstuffs,* Jemmali, M., Ed., INSERM-INRA, Paris, 1977, 1765.
3. **Drobotko, V. G.,** Novoe gribkovoe zabolevanie loschadej i lidei (stachibotriotoxicosis). New fungal disease of horses and people (stachibotriotoxicosis) (transl.), Kiev Academia of Ukrainian Soviet Socialist Republic, 1949, 37.
4. **Linnik, F. A.,** Materiali po isucheniu stachibotriotoxicosa u ludei, Materials on the study of stachibotriotoxicosis in humans (transl.), Kiev Academia of Ukrainian Soviet Socialist Republic, 1949, 27.
5. **Sarkisov, A. C.,** *Mycoses,* State Publishing House for Agricultural Literature, Moscow, 1954, 216.
6. **Bilai, V. I and Pidoplicko, N. M.,** Fuzarii, Naukova Dumka, 1977, 360.
7. **Leonov, A. N.,** Current view of the chemical nature of factors responsible for alimentary toxic aleukia, in *Mycotoxins in Human and Animal Health,* Rodricks, J. V., Hesseltine, C. W., and Mehlman, M. A., Eds., Pathotox Publications Inc., Illinois, 1977, 323.
8. **Uraguchi, K.,** Yellowed rice toxins: citreoviridin, in *Microbial Toxins,* Ciegler, A., Kadis, S., and Ajl, S. J., Eds., Academic Press, New York, 1971, 367.
9. **Krogh, P.,** Mycotoxic porcine nephropathy: a possible model for Balkan endemic nephropathy, in *Endemic Nephropathy,* Puchlev, A., Ed., Bulgarian Academy of Sciences, 1974, 266.
10. **Elling, F. and Moller, T.,** Mycotoxic nephropathy in pigs, *Bull. W.H.O.,* 49, 411, 1973.
11. **Dotchev, V. D.,** The endemic (Balkan) nephropathy in Bulgaria, *Munch. Med. Wochenschr.,* 115, 537, 1973.
12. **Hrabar, A., Šujaga, K., Borčič, B., Aleraj, B., Čeovič, S., and Cvoriščec, D.,** Morbidity and mortality from endemic nephropathy in the village of Kaniže, *Arch. Hig. Rada,* 27, 137, 1976.
13. **Chernozemsky, I. N., Stoyanov, I. S., Petkova-Bocharova, T. K., Nicolov, I. G., Graganov, I. V., Stoichev, I. I., Tanchev, Y., Naidenov, D., and Kalcheva, W. D.,** Geographic correlation between occurrence of endemic nephropathy and urinary tract tumours in Vratze district, Bulgaria, *Int. J. Cancer,* 19, 1, 1977.
14. **Heptinstall, R. H.,** *Pathology of the Kidney,* Heptinstall, R. H., Ed., Churchill Livingstone, London, 1966, 457.
15. **Hult, K., Plestina, R., Ceovic, S., Habazin-Nowak, V., and Radic, B.,** Ochratoxin A in human blood: analytical results and confirmational tests from a study in connection with Balkan endemic nephropathy, presented at Int. IUPAC Symp. Mycotoxins and Phycotoxins, Vienna, August 30 to September 1, 1982.
16. **Krogh, P.,** Causal association of mycotoxic nephropathy, *Acta Pathol. Microbiol. Scand. Sect. A.,* 269, 1, 1978.
17. **Blount, W. P.,** Turkey X disease, *J. Br. Turkey Fed.,* 9, 55, 1961.

18. **Loosmore, R. M. and Markson, L. M.,** Poisoning of cattle by Brazilian groundnut meal, *Vet. Rec.,* 73, 813, 1961.
19. **Loosmore, R. M. and Harding, J. D. J.,** A toxic factor in Brazilian groundnut causing liver damage in pigs, *Vet. Rec.,* 73, 1362, 1961.
20. **Sargeant, K. D., Kelly, J., Carnaghan, R. B. A., and Allcroft, R.,** The assay of a toxic principle in certain groundnut meals, *Vet. Rec.,* 73, 1219, 1961.
21. **Van der Zijden, A. S. M., Blanche Koelensmid, W. A. A., Boldingh, J., Barrett, C. B., Ord, W. O., and Philp, J.,** *Aspergillus flavus* and turkey X disease, *Nature (London),* 195, 1060, 1962.
22. **Nesbitt, B. F., O'Kelly, J., Sargeant, K., and Sheridan, A.,** Toxic metabolites of *Aspergillus flavus, Nature (London),* 195, 1062, 1962.
23. **Asao, T., Buchi, G., Abdel-Kader, M., Chang, S. B., Wick, E., and Wogan, G. N.,** Aflatoxins B and G, *J. Am. Chem. Soc.,* 85, 1706, 1963.
24. **Asao, T., Buchi, G., Abdel-Kader, M., Chang, S. B., Wick, E., and Wogan, G. N.,** The structures of aflatoxins B and G, *J. Am. Chem. Soc.,* 87, 882, 1965.
25. **Chang, S. B., Abdel-Kader, M., Wick, E., and Wogan, G. N.,** Aflatoxin B_2: chemical identity and biological activity, *Science,* 142, 1191, 1963.
26. **Hartley, R. D., Nesbitt, B. F., and O'Kelly, J.,** Toxic metabolites of *Aspergillus flavus, Nature (London),* 198, 1056, 1963.
27. **Allcroft, R. and Carnaghan, R. B. A.,** Groundnut toxicity: an examination for toxin in human food products from animals fed toxic groundnut meal, *Vet. Rec.,* 75, 259, 1963.
28. **Allcroft, R., Rogers, H., Lewis, G., Nabney, B., and Best, P. E.,** Metabolism of aflatoxin in sheep: excretion of the "milk-toxin", *Nature (London),* 209, 154, 1966.
29. **Hesseltine, C. W., Sorenson, W. G., and Smith, M.,** Taxonomic studies of the aflatoxin producing strains in the *Aspergillus flavus* group, *Mycologia,* 62, 123, 1970.
30. **Waltking, A. E.,** Fate of aflatoxin during roasting and storage of contaminated peanut products, *J. Assoc. Off. Anal. Chem.,* 54, 533, 1971.
31. **Stephenson, L. W. and Russell, T. E.,** The association of *Aspergillus flavus* with hemipterons and other insects infesting cotton bracts and foliage, *Phytopathology,* 64, 1502, 1974.
32. **Stoloff, L.,** Occurrence of mycotoxins in foods and feeds, in *Mycotoxins and Other Fungal Related Food Problems,* Rodricks, J. V., Ed., American Chemical Society, Washington, D.C., 1976, 23.
33. **Shank, R. C. Wogan, G. N., Gibson, J. B., and Nondasuta, A.,** Dietary aflatoxins and human liver cancer. II. Aflatoxins in market foods and foodstuffs of Thailand and Hong Kong, *Food Cosmet. Toxicol.,* 10, 61, 1972.
34. **Shotwell, O. L., Hesseltine, C. W., Burmeister, H. R., Kwolek, W. F., Shannon, G. M., and Hall, H. H.,** Survey of cereal grains and soybeans for the presence of aflatoxin: wheat, grain sorghum, and oats, *Cereal Chem.,* 46, 446, 1969.
35. **Seenappa, M., Stobbs, L. W., and Kempton, A. G.,** The role of insects in the biodeterioration of Indian red peppers by fungi, *Int. Biodeterior. Bull.,* 15, 96, 1979.
36. **Buchanan, J. R., Sommer, N. F., and Fortlage, R. J.,** *Aspergillus flavus* infection and aflatoxin production in fig fruits, *Appl. Microbiol.,* 30, 238, 1975.
37. **Kiermeier, R.,** The significance of aflatoxin in the dairy industry, *Ann. Bull. Int. Dairy Fed.,* 38, 1, 25, 1977.
38. **Fremy, J. M., Cariou, T., and Bonnet, C.,** Natural occurrence of aflatoxin M_1 in milk and whey powders in France, presented at Int. IUPAC Symp. Mycotoxins and Phycotoxins, Vienna, August 30 to September 1, 1982.
39. **Van Egmond, H. P., Paulsch, W. E., Sizoo, E. A., and Schuller, P. L.,** Occurrence of aflatoxin M_1 in Dutch consumption milk, presented at Int. IUPAC Symp. Mycotoxins and Phycotoxins, Vienna, August 30 to September 1, 1982.
40. **Van Zytweld, W. A.,** Aflatoxicosis: The Presence of Aflatoxins or Their Metabolites in Livers and Skeletal Muscle of Chickens, Master's Dissertation, Kansas State University, Manhattan, 1968.
41. **Armbrecht, B. H.,** Aflatoxin residues in food and feed derived from plant and animal sources, *Residue Rev.,* 41, 13, 1971.
42. **Krogh, P.,** Residues of aflatoxin in swine given aflatoxin containing feed, in Agricultural Res. Sta. Yearbook, R. Vet. Agric. Coll., Copenhagen, 1970, 84.
43. **Furtado, R. M., Pearson, A. M., Hogberg, M. G., and Miller, E. R.,** Aflatoxin residues in the tissue of pigs fed a contaminated diet, *J. Agric. Food Chem.,* 27, 1351, 1979.
44. **Jacobson, W. C. and Wiseman, H. G.,** The transmission of aflatoxin B_1 into eggs, *Poult. Sci.,* 53, 1743, 1974.
45. **Trucksess, M. W., Stoloff, L., Young, K., Wyatt, R. D., and Miller, B. L.,** Aflatoxicol and aflatoxins B_1 and M_1 in eggs and tissues of laying hens consuming aflatoxin-contaminated feed, *Poult. Sci.,* 62, 2176, 1983.
46. **Carnaghan, R. B. A. Hartley, R. D., and O'Kelly, J.,** Toxicity and fluorescence properties of aflatoxins, *Nature (London),* 200, 1101, 1963.
47. **Burmeister, H. R. and Hesseltine, C. W.,** Survey of the sensitivity of microorganisms to aflatoxins, *Appl. Microbiol.,* 14, 403, 1966.

48. **Verrett, M. J., Winbush, J., Reynaldo, E. F., and Scott, W. F.,** Collaborative study of the chicken embryo bioassay for aflatoxin B_1, *J. Assoc. Off. Anal. Chem.,* 56, 901, 1973.
49. **Horwitz, W., Senzel, A., Reynolds, H., and Park, D. L.,** Natural poisons, in *Official Methods of Analysis of the Association of Official Analytical Chemists,* Washington, D.C., 1975, 24.
50. **Hsieh, D. P. H., Fitzell, D. L., Miller, J. L., and Seiber, J. N.,** High-pressure liquid chromatography of oxidative aflatoxin metabolites, *J. Chromatogr.,* 117, 474, 1976.
51. **Panalaks, T. and Scott, P. M.,** Sensitive silica gel-packed flowcell for fluorometric detection of aflatoxins by high pressure liquid chromatography, *J. Assoc. Off. Anal. Chem.,* 60, 583, 1977.
52. **Langone, J. J. and VanVunakis, H.,** Aflatoxin B_1: specific antibodies and their use in radioimmunoassay, *J. Natl. Cancer Inst.,* 56, 591, 1976.
53. **Chu, F. S. and Ueno, I.,** Production of antibody against aflatoxin B_1, *Appl. Environ. Microbiol.,* 33, 1125, 1977.
54. **Lawellin, D. W., Grant, D. W., and Joyce, B. K.,** Enzyme-linked immunosorbent analysis of aflatoxin B_1, *Appl. Environ. Microbiol.,* 34, 94, 1977.
55. **Pestka, J. J., Gaur, P. K., and Chu, F. S.,** Quantitation of aflatoxin B_1 and aflatoxin B_1 antibody by an enzyme-linked immunosorbent microassay, *Appl. Environ. Microbiol.,* 40, 1027, 1980.
56. **Sizaret, P., Malaveille, C., Montesano, R., and Frayssinet, C.,** Detection of aflatoxins and related metabolites by radioimmunoassay, *J. Natl. Cancer Inst.,* 69, 1375, 1982.
57. **Allcroft, R.,** Aflatoxicosis in farm animals, in *Aflatoxin: Scientific Backround, Control, and Implications,* Goldblatt, L. A., Ed., Academic Press, New York, 1969, 237.
58. **Newberne, P. M. and Butler, W. H.,** Acute and chronic effects of aflatoxin on the liver of domestic and laboratory animals. A review, *Cancer Res.,* 29, 236, 1969.
59. **Butler, W. H.,** Aflatoxicosis in laboratory animals, in *Aflatoxin: Scientific Backround, Control, and Implications,* Goldblatt, L. A., Ed., Academic Press, New York, 1969, 223.
60. **Tupule, P. G., Madhaven T. V., and Gopalan, C.,** Effect of feeding aflatoxin to young monkeys, *Lancet,* 1, 962, 1964.
61. **Butler, W. H.,** Acute toxicity of aflatoxin B_1 in rats, *Br. J. Cancer,* 18, 756, 1964.
62. **Gabliks, J. W., Schaeffer, W., Friedman, L., and Wogan, G. N.,** Effect of aflatoxin B_1 on cell cultures, *J. Bacteriol.,* 90, 720, 1965.
63. **Legator, M. S., Zuffante, S. M., and Harp, A. R.,** Aflatoxin: effect on cultured heteroploid human embryonic lung cells, *Nature (London),* 208, 345, 1965.
64. **Sullman, S. F., Armstrong, S. J., Zuckerman, A. Z., and Rees, K. R.,** Further studies on the toxicity of the aflatoxins on human cell cultures, *Br. J. Exp. Pathol.,* 51, 314, 1970.
65. **Lancaster, M. C., Jenkins, F. P., and Philip, J. M.,** Toxicity associated with certain samples of groundnuts, *Nature (London),* 192, 1095, 1961.
66. **Wogan, G. N., Paglialunga, S., and Newberne, P. M.,** Carcinogenic effects of low dietary levels of aflatoxin B_1 in rats, *Food Cosmet. Toxicol.,* 12, 681, 1974.
67. **Sinnhuber, R. O., Wales, J. H., Ayres, J. L., Engebrecht, R. H., and Amend, D. L.,** Dietary factors and hepatoma in rainbow trout (*Salmo gairdneri*). Aflatoxins in vegetable protein feedstuffs, *J. Natl. Cancer Inst.,* 4, 711, 1968.
68. **Carnaghan, R. B. A.,** Hepatic tumors in ducks fed a low level of toxic groundnut meal, *Nature (London),* 208, 308, 1965.
69. **Adamson, R. H., Correa, P., and Dalgard, D. W.,** Occurrence of a primary liver carcinoma in a rhesus monkey fed aflatoxin B_1, *J. Natl. Cancer Inst.,* 50, 549, 1973.
70. **Gopalan, C., Tulpule, P. G., and Krisnamurthi, D.,** Induction of hepatic carcinoma with aflatoxin in the rhesus monkey, *Food Cosmet. Toxicol.,* 10, 519, 1972.
71. **Tilak, T. B. G.,** Induction of cholangiocarcinoma following treatment of a rhesus monkey with aflatoxin, *Food Cosmet. Toxicol.,* 13, 247, 1975.
72. **Reddy, J. K. and Svoboda, D. J.,** Aflatoxin B_1-induced liver tumors in *Tupaia glis* (tree shrews), a nonhuman primate, *Fed. Proc.,* 34, 827, 1975.
73. **Sieber, S. M., Correa, P., Dalgard, D. W., and Adamson, R. H.,** Induction of osteogenic sarcomas and tumors of the hepatobiliary system in nonhuman primates with aflatoxin B_1, *Cancer Res.,* 39, 4545, 1979.
74. **Butler, W. H. and Barnes, J. M.,** Carcinoma of the glandular stomach in rats given diets containing aflatoxin, *Nature (London),* 209, 90, 1966.
75. **Wogan, G. N. and Newberne, P. M.,** Dose-response characteristics of aflatoxin B_1 carcinogenesis in the rat, *Cancer Res.,* 27, 2370, 1967.
76. **Epstein, S. M., Bartus, B., and Farber, E.,** Renal epithelial neoplasms induced in male Wistar rats by oral aflatoxin B_1, *Cancer Res.,* 29, 1045, 1969.
77. **Ward, J. M., Sontag, J. M., Weiburger, E. K., and Brown, C. A.,** Effect of lifetime exposure to aflatoxin B_1 in rats, *J. Natl. Cancer Inst.,* 55, 107, 1975.
78. **Butler, W. H., Greenblatt, M., and Lijinsky, W.,** Carcinogenesis in rats by aflatoxin B_1, G_1, and B_2, *Cancer Res.,* 29, 2206, 1969.

79. Canton, J. H., Kroes, R., Van Logten, M. J., Van Schothorst, M., Stavemitter, J. F. C., and Verhulsdonk, C. A. H., The carcinogenicity of aflatoxin M_1 in rainbow trout, *Food Cosmet. Toxicol.*, 13, 441, 1975.
80. Sinnhuber, R. O., Lee, D. J., Wales, J. H., Landers, M. K., and Keyl, A. C., Hepatic carcinogenesis by aflatoxin M_1 in rainbow trout (*Salmo gairdneri*) and its enhancement by cyclopropene fatty acids, *J. Natl. Cancer Inst.*, 53, 1285, 1974.
81. Wogan, G. N. and Paplialunga, S., Carcinogenicity of synthetic aflatoxin M_1 in rats, *Food Cosmet. Toxicol.*, 12, 381, 1974.
82. Wogan, G. N., Edwards, G. S., and Newberne, P. M., Structure-activity relationship in toxicity and carcinogenicity of aflatoxins and analogs, *Cancer Res.*, 31, 1936, 1971.
83. Butler, W. H. and Wigglesworth, J. S., The effects of aflatoxin B_1 on the pregnant rat, *Br. J. Pathol.*, 47, 242, 1966.
84. Butler, W. H., The effect of maternal liver injury and dietary reduction of foetal growth in the rat, *Food Cosmet. Toxicol.*, 9, 57, 1971.
85. Elis, J. and DiPaolo, J. A., Aflatoxin B_1-induction of malformations, *Arch. Pathol.*, 83, 53, 1967.
86. DiPaolo, J. A., Elis, J., and Erwin, H., Teratogenic responses by hamsters, rats and mice to aflatoxin B_1, *Nature (London)*, 215, 638, 1967.
87. Verrett, M. J., Marliac, J. P., and McLaughlin, J., Jr., Use of the chick embryo in the assay of aflatoxin toxicity, *J. Assoc. Off. Anal. Chem.*, 47, 1003, 1964.
88. Bassier, O. and Adenkule, H., Teratogenic action of aflatoxin B_1, palmotoxin Bo and palmotoxin Go on the chick embryo, *J. Pathol.*, 102, 49, 1970.
89. Stoloff, M. J., Verrett, M. J., Dontzman, J., and Reynaldo, E. F., Toxicological study of aflatoxin P_1 using the fertile chicken egg, *Toxicol. Appl. Pharmacol.*, 23, 528, 1972.
90. Hintz, H. F., Heitman, H. J., Booth, A. N., and Gangne, W. E., Effects of aflatoxin on reproduction in swine, *Proc. Soc. Exp. Biol. Med.*, 126, 144, 1967.
91. Lilly, L. J., Induction of chromosome aberrations by aflatoxin, *Nature (London)*, 207, 433, 1965.
92. Withers, R. F. J., The action of some lactones and related compounds on human chromosomes, in *Proceedings of the Symposium Mutational Process*, Lauda, Z., Ed., Czechoslovak Academy of Sciences, Prague, 1966, 359.
93. Dolimpio, D. A., Green, S., Legator, M., and Jacobson, C. B., Effect of aflatoxin on human leucocytes, *Proc. Soc. Exp. Biol. Med.*, 127, 1968.
94. Maher, V. M. and Summers, W. C., Mutagenic action of aflatoxin B_1 on transforming DNA and inhibition of DNA template activity *in vitro*, *Nature (London)*, 225, 68, 1970.
95. Ong, T., Mutagenic activities of aflatoxin B_1 and G_1 in *Neurospora crassa*, *Mol. Gen. Genet.*, 111, 159, 1971.
96. Schimmer, O. and Werner, R., Mutagenic effect of aflatoxin B_1 on nuclear and extranuclear RNA in *Chlamydommas reinhardii*, *Mutat. Res.*, 26, 423, 1974.
97. Lamb, M. J. and Lilly, L. J., Induction of recessive lethals in *Drosophila melanogaster* by aflatoxin B_1, *Mutat. Res.*, 11, 430, 1971.
98. Garner, R. C., Miller, E. C., Miller, J. A., Garner, J. V., and Hanson, R. S., Formation of a factor lethal for *S. typhimurium* TA 1530 and TA 1531 on incubation of aflatoxin B_1 with rat liver microsomes, *Biochem. Biophys. Res. Commun.*, 45, 774, 1971.
99. Ames, B. N., Durston, W. E., Yamasaki, E., and Lee, F. D., Carcinogens are mutagens: a simple test system combining liver homogenates for activation and bacteria for detection, *Proc. Natl. Acad. Sci. U.S.A.*, 70, 2281, 1973.
100. Garner, R. C., Miller, E. C., and Miller, J. A., Liver microsomal metabolism of aflatoxin B_1 to a reactive derivate toxic to *Salmonella typhimurium* TA 1530, *Cancer Res.*, 32, 2058, 1972.
101. Garner, R. C. and Wright, C. M., Induction of mutations in DNA-repair deficient bacteria by a liver microsomal metabolite of aflatoxin B_1, *Br. J. Cancer*, 28, 544, 1973.
102. Wong, J. J. and Hsieh, D. P. H., Mutagenicity of aflatoxins related to their metabolism and carcinogenic potential, *Proc. Natl. Acad. Sci. U.S.A.*, 73, 2241, 1976.
103. Garner, R. C., Chemical evidence for the formation of a reactive aflatoxin B_1 metabolite by hamster liver microsomes, *FEBS Lett.*, 36, 261, 1973.
104. Swenson, D. H., Miller, E. C., and Miller, J. A., Aflatoxin B_1-2,3-oxide: evidence for its formation in rat liver *in vivo* and by human liver microsomes *in vitro*, *Biochem. Biophys. Res. Commun.*, 60, 1036, 1974.
105. Boonchuvit, B. and Hamilton, P. B., Interaction of aflatoxin and paratyphoid infections in broiler chickens, *Poult. Sci.*, 54, 1567, 1975.
106. Hamilton, P. B. and Harris, J. R., Interaction of aflatoxins with *Candida albicans* infections and other stresses in chickens, *Poult. Sci.*, 50, 906, 1971.
107. Richard, J. L., Pier, A. C., Cysewski, S. J., and Graham, C. K., Effect of aflatoxin and aspergillosis on turkey poults, *Avian Dis.*, 17, 111, 1973.
108. Edds, G. T. and Simpson, C. F., Cecal coccidiosis in poultry as affected by prior exposure to aflatoxin B_1, *Am. J. Vet. Res.*, 37, 65, 1976.

109. **Edds, G. T., Nair, K. P. C., and Simpson, C. F.,** Effect of aflatoxin B_1 on resistance in poultry against cecal coccidiosis and Marek's disease, *Am. J. Vet. Res.,* 34, 819, 1973.
110. **Thaxton, J. P., Tung, H. T., and Hamilton, P. B.,** Immunosuppression in chickens by aflatoxin, *Poult. Sci.,* 53, 721, 1974.
111. **Richard, J. L. and Thurston, J. R.,** Effect of aflatoxin on phagocytosis of *Aspergillus fumigatus* spores by rabbit alveolar macrophages, *Appl. Microbiol.,* 30, 44, 1975.
112. **Michael, G. Y., Thaxton, P., and Hamilton, P. B.,** Impairment of the reticuloendothelial system of chickens during aflatoxicosis, *Poult. Sci.,* 52, 1206, 1973.
113. **Mohapatra, N. and Roberts, J. F.,** Effects of aflatoxin B_1 on rat peritoneal macrophages and mouse fibroblasts (L — M cells), *Gen. Pharmacol.,* 10, 471, 1979.
114. **Chang, C. F. and Hamilton, P. B.,** Impaired phagocytosis by heterophils from chickens during aflatoxicosis, *Toxicol. Appl. Pharmacol.,* 48, 459, 1979.
115. **Thurston, J. R., Baetz, A. L., Cheville, N. F., and Richard, J. L.,** Acute aflatoxicosis in guinea pigs: sequential changes in serum proteins, complement, C_4 and liver enzymes and histopathologic changes, *Am. J. Vet. Res.,* 41, 1272, 1980.
116. **Hahon, N., Booth, J. A., and Stewart, J. D.,** Aflatoxin inhibition of viral interferon induction, *Antimicrob. Agents Chemother.,* 16, 277, 1979.
117. **Butler, W. H. and Cliford, J. I.,** Extraction of aflatoxin from rat liver, *Nature (London),* 206, 1045, 1965.
118. **Pong, R. S. and Wogan, G. N.,** Toxicity, biochemical and fine structural effects of synthetic aflatoxins M_1 and B_1, *J. Natl. Cancer Inst.,* 47, 585, 1971.
119. **Purchase, I. F. H.,** Aflatoxin residues in food of animal origin, *Food Cosmet. Toxicol.,* 10, 531, 1972.
120. **Masri, M., Booth, N., and Hsieh, P. H.,** Comparative metabolic conversion of aflatoxin B_1 to M_1 and Q_1 by monkey, rat and chicken liver, *Life Sci.,* 15, 203, 1974.
121. **Patterson, D. S. P. and Roberts, B. A.,** The formation of aflatoxin B_2a and G_2a and their degradation products during the *in vitro* detoxification of aflatoxin by liver of certain avian and mammalian species, *Food Cosmet. Toxicol.,* 8, 527, 1970.
122. **Patterson, D. S. P.,** Metabolism as a factor in determining the toxic action of the aflatoxin in different animal species, *Food Cosmet. Toxicol.,* 11, 287, 1973.
123. **Detroy, R. W. and Hesseltine, C. W.,** Aflatoxicol: structure of a new transformation product of aflatoxin B_1, *Can. J. Biochem.,* 48, 830, 1970.
124. **Patterson, D. S. P. and Roberts, B. A.,** Steroid sex hormones as inhibitors of aflatoxin metabolism in liver homogenates, *Experientia (Basel),* 28, 929, 1972.
125. **Patterson, D. S. P. and Roberts, B. A.,** Aflatoxin metabolism in duck-liver homogenates: the relative importance of reversible cyclopentenone reduction and hemiacetal formation, *Food Cosmet. Toxicol.,* 10, 501, 1972.
126. **Dalezios, J. I., Wogan, G. N., and Weinreb, S. M.,** Aflatoxin P_1: a new aflatoxin metabolite in monkeys, *Science,* 171, 584, 1971.
127. **Dalezios, J. I. and Wogan, G. N.,** Metabolism of aflatoxin B_1 in Rhesus monkeys, *Cancer Res.,* 32, 2297, 1972.
128. **Buchi, G. H., Muller, P. M., Roebuck, B. D., and Wogan, G. N.,** Aflatoxin Q_1. A major metabolite of aflatoxin B_1 produced by human liver, *Res. Commun. Chem. Pathol. Pharmacol.,* 8, 585, 1974.
129. **Masri, M. S., Haddon, W. F., Lundin, R. E., and Hsieh, D. P. H.,** Aflatoxin Q_1: a newly identified major metabolite of aflatoxin B_1 in monkey liver, *J. Agricul. Food Chem.,* 22, 512, 1974.
130. **Hsieh, D. P. H., Salhab, A. S., Wong, J. J., and Yang, S. L.,** Toxicity of aflatoxin Q_1 as evaluated with the chicken embryo and bacterial auxotrophs, *Toxicol. Appl. Pharmacol.,* 30, 237, 1974.
131. **Essigmann, J. M., Croy, R. G., Nadzan, A. M., Busby, W. F., Jr., Reinhold, V. N., Buchi, G., and Wogan, G. N.,** Structural identification of the major DNA adduct formed by aflatoxin B_1 *in vitro, Proc. Natl. Acad. Sci. U.S.A.,* 74, 1870, 1977.
132. **Miller, E. C. and Miller, J. A.,** Biochemical mechanisms of chemical carcinogenesis, in *Molecular Biology of Cancer,* Busch, H., Ed., Academic Press, New York, 1974, 377.
133. **Godoy, H. M. and Neal, G. E.,** Some studies of the effects of aflatoxin B_1 *in vivo* and *in vitro* on nucleic acid synthesis in rat and mouse, *Chem. Biol. Interact.,* 13, 257, 1976.
134. **Campbell, T. C. and Hayes, J. R.,** The role of aflatoxin metabolism and its toxic lesion, *Toxicol. Appl. Pharmacol.,* 35, 199, 1976.
135. **Hsieh, D. P. H., Wong, Z. A., Wong, J. J., Michas, C., and Ruebner, B. H.,** Comparative metabolism of aflatoxin, in *Mycotoxins in Human and Animal Health,* Rodricks, J. V., Hesseltine, C. W., and Mehlman, M. A., Eds., Pathotox Publications, Park Forest South, IL, 1977, 37.
136. **Pong, R. S. and Wogan, G. N.,** Time course and dose-response characteristics of aflatoxin B_1 effects on rat liver RNA polymerase and ultrastructure, *Cancer Res.,* 30, 294, 1970.
137. **Clifford, J. I. and Rees, K. R.,** The action of aflatoxin B_1 on the rat liver, *Biochem. J.,* 102, 65, 1967.
138. **Seegers, J. C. and Pitout, M. J.,** DNA repair synthesis in all cell cultures exposed to low doses of aflatoxin B_1 or sterigmatocystin, *S. Afr. Lab. Clin. Med.,* 19, 961, 1973.

139. **Neal, G. E. and Godoy, H. M.,** The effect of pre-treatment with phenobarbitone on the activation of aflatoxin B_1 by rat liver, *Chem. Biol. Interact.,* 14, 279, 1976.
140. **Doherty, W. P. and Campbell, T. C.,** Inhibition of rat-liver mitochondria electrontransport flow by aflatoxin B_1, *Res. Commun. Chem. Pathol. Pharmacol.,* 3, 601, 1972.
141. **Wei, R. D., Lee, S. S., Liu, G. X., and Hsu, C. M.,** Effect of aflatoxin B_1 on the biosynthesis of lipids in the rat, *Chin. J. Physiol.,* 20, 131, 1968.
142. **Kato, R., Onoda, K., and Omori, Y.,** Effect of aflatoxin B_1 on the incorporation of ^{14}C-acetate into cholesterol by rat liver, *Experientia,* 25, 1026, 1969.
143. **Williams, D. J. and Rabin, B. R.,** Disruption by carcinogens of the hormone dependent association of membranes with polysomes, *Nature (London),* 232, 102, 1971.
144. **Pokrovsky, A. A., Kravchenko, L. V., and Tutelyan, V. A.,** Effect of aflatoxin on rat liver lysosomes, *Toxicon,* 10, 25, 1972.
145. **Tung, H. T., Donaldson, W. E., and Hamilton, P. B.,** Effects of aflatoxin on some marker enzymes of lysosomes, *Biochim. Biophys. Acta,* 22, 665, 1970.
146. **Shankaran, R., Raj, H. G., and Vankitasubramanian, T. A.,** Effect of aflatoxin on carbohydrate metabolism in chick liver, *Enzymology,* 39, 371, 1970.
147. **Newberne, P. M. and Wogan, G. N.,** Sequential morphologic changes in aflatoxin B_1 carcinogenesis in the rat, *Cancer Res.,* 28, 770, 1968.
148. **Newberne, P. M. and Williams, G.,** Inhibition of aflatoxin carcinogenesis by diethylstilbestrol in male rats, *Arch. Environ. Health,* 19, 489, 1969.
149. **Goodall, C. M. and Butler, W. H.,** Aflatoxin carcinogenesis. Inhibition of liver cancer induction in hypophysectomized rats, *Int. J. Cancer,* 4, 422, 1969.
150. **Chedid, A., Bundeally, A. E., and Mendenhall, C. L.,** Inhibition of hepatocarcinogenesis by adrenocorticotropin in aflatoxin B_1 – treated rats, *J. Natl. Cancer Inst.,* 58, 339, 1977.
151. **Madhavan, T. V. and Gopalan, C.,** The effect of dietary protein on carcinogenesis of aflatoxin, *Arch. Pathol.,* 85, 133, 1968.
152. **McLean, A. E. M. and McLean, E. K.,** Protein depletion and toxic liver injury due to carbon tetrachloride and aflatoxin, *Proc. Nutr. Soc. Engl. Scol.,* 26, 13, 1967.
153. **Newberne, P. M., Rogers, A. E., and Wogan, G. N.,** Hepatorenal lesions in rats fed a low lipotrope diet and exposed to aflatoxin, *J. Nutr.,* 94, 331, 1968.
154. **Rogers, A. E. and Newberne, P. M.,** Aflatoxin B_1 carcinogenesis in lipotrope deficient rats, *Cancer Res.,* 29, 1965, 1969.
155. **Rogers, A. E. and Newberne, P. M.,** Diet and aflatoxin B_1 toxicity in rats, *Toxicol. Appl. Pharmacol.,* 20, 113, 1971.
156. **Rogers, A. E. and Newberne, P. M.,** Nutrition and aflatoxin carcinogenesis, *Nature (London),* 229, 62, 1971.
157. **Temcharoen, P., Anukarahanonta, T., and Bhamarapravati, N.,** Influence of dietary protein and vitamin B_{12} on the toxicity and carcinogenicity of aflatoxins in rat liver, *Cancer Res.,* 38, 2185, 1978.
158. **Reddy, G. S., Tilak, T. B., and Krishnamuthi, D.,** Susceptibility of vitamin A-deficient rats to aflatoxin, *Food Cosmet. Toxicol.,* 11, 467, 1973.
159. **Newberne, P. M. and Rogers, A. E.,** Rat colon carcinomas associated with aflatoxin and marginal vitamin A, *J. Natl. Cancer Inst.,* 50, 439, 1973.
160. **Newberne, P. M., Weigert, J., and Kula, N.,** Effects of dietary fat on hepatic mixed function oxidases and hepatocellular carcinoma induced by aflatoxin B_1 in rats, *Cancer Res.,* 39, 3986, 1979.
161. **Newberne, P. M. and Conner, M. W.,** Effect of selenium on acute response to aflatoxin B_1, in *A Symposium. Trace Substances in Environmental Health — VIII,* Hemphill, D. D., Ed., University of Missouri, Columbia, 1974, 323.
162. **Davila, J. C., Edds, G. T., Osuna, O., and Simpson, C. F.,** Modification of the effects of aflatoxin B_1 and warfarin in young pigs given selenium, *Am. J. Vet. Res.,* 44, 1877, 1983.
163. **Gregory, J. F. and Edds, G. T.,** Effect of dietary selenium on the metabolism of aflatoxin B_1 in turkeys, *Food Chem. Toxicol.,* 22, 637, 1984.
164. **Aleksandrowicz, J., Dobrowolski, J., and Lisiewicz, J.,** Effect of selenium on immunosuppressive and teratogenic properties of aflatoxin B_1, *Rev. Esp. Oncol.,* 22, 239, 1975.
165. **Osuna, O., Edds, G. T., and Simpson, Ch. F.,** Toxicology of aflatoxin B_1, warfarin and cadmium in young pigs: metal residues and pathology, *Am. J. Vet. Res.,* 43, 1395, 1982.
166. **Dvořáčková, I.,** Aflatoxin inhalation and alveolar cell carcinoma, *Br. Med. J.,* 1, 691, 1976.
167. **Baxter, C. S., Wey, H. E., and Burg, W. R.,** A prospective analysis of the potential risk associated with inhalation of aflatoxin-contaminated grain dusts, *Food Cosmet. Toxicol.,* 19, 765, 1981.
168. **Sorenson, W. G., Simpson, J. P., Peach, M. J., Thedell, T. D., and Olenchock, S. A.,** Aflatoxin in respirable corn dust particles, *J. Toxicol. Environ. Health,* 7, 669, 1981.
169. **Hayes, R. B., Van Nieuwenhuize, J. P., Raatgever, J. W., and Kate, F. J. W.,** Aflatoxin exposures in the industrial setting: an epidemiological study of mortality, *Food Chem. Toxicol.,* 22, 39, 1984.

170. **Riley, R. T., Kemppainen, B. W., and Norred, W. P.,** Penetration of aflatoxins through isolated human epidermis, *J. Toxicol. Environ. Health,* 15, 769, 1985.
171. **Ling, K. H., Wang, J. J., Wu, R., Tung, T. C., Lin, C. K., Lin, S. S., and Lin, T. M.,** Intoxication possibly caused by aflatoxin B_1 in mouldy rice in Shuang-Chih township, *J. Formosan Med. Assoc.,* 66, 517, 1967.
172. **Van Walbeck, W., Scott, P. M., and Thatcher, F. S.,** Mycotoxins from food-borne fungi, *Can. J. Microbiol.,* 14, 131, 1968.
173. **Serck-Hansen, A.,** aflatoxin-induced fatal hepatitis? A case report from Uganda, *Arch. Environ. Health,* 20, 729, 1970.
174. **Bosenberg, H.,** Aflatoxinwirkung bei Tier und Mensch, *Z. Lebensm. Untersuch. Forsch.,* 151, 245, 1972.
175. **Krishnamachari, K. A. V. R., Bhat, R. V., Nagarajan, V., and Tilak, T. B. G.,** Hepatitis due to aflatoxicosis, *Lancet,* 1, 1061, 1975.
176. **Tandon, B. N., Krishnamurthy, L., Kosky, A., Tandon, H. D., Ramalingaswami, V., Bhandari, J. R., Mathur, M. M., and Mathur, P. D.,** Study of an epidemic of jaundice, presumably due to toxic hepatitis, in northwest India, *Gastroenterology,* 72, 488, 1977.
177. **Tandon, H. D. and Tandon, B. N.,** Clinical and pathological study of an human outbreak of aflatoxicosis, presented at Int. Symp. on Mycotoxins, Cairo, September 6 to 8, 1981.
178. **Ngindu, A., Kenya, P. R., Ocheng, D. M., Omondi, T. N., Ngare, W., Gatei, D., Johnson, B. K., Ngira, J. A., Nandova, H., Jansen, A. J., Kavita, J. N., and Siongok, T. A.,** Outbreak of acute hepatitis caused by aflatoxin poisoning in Kenya, *Lancet,* 1, 1346, 1982.
179. **Hendrickse, R. G., Coulter, J. B. S., Lamplugh, S. M., MacFarlane, S. B. H., Williams, T. E., Omer, M. I. A., and Suliman, G. I.,** Aflatoxins and kwashiorkor: a study in Sudanese children, *Br. Med. J.,* 285, 843, 1982.
180. **Hendrickse, R. G.,** The influence of aflatoxins on child health in the tropics with particular reference to kwashiorkor, *Trans. R. Soc. Trop. Med. Hyg.,* 78, 427, 1984.
181. **Long, D. A.,** Aflatoxins and kwashiorkor, *Br. Med. J.,* 285, 1208, 1982.
182. **Amla, I., Kamala C. S., Gopalakrishma, G. S., Jayraj, A. P., Sreenivasamurthy, V., and Parpia, H. A. B.,** Cirrhosis in children from peanut meal contaminated by aflatoxin, *Am. J. Clin. Nutr.,* 24, 609, 1971.
183. **Payet, M., Gros, J., Quenum, C., Sankale, M., and Moulanier, M.,** Deux observations d'enfants ayant consommé de facon prolongée des farines souilléss par *"Aspergillus flavus", Presse Med.,* 13, 649, 1966.
184. **Robinson, P.,** Infantile cirrhosis of the liver in India. With special reference to probable aflatoxin etiology, *Clin. Pediatr.,* 6, 57, 1967.
185. **Alpert, M. E., Hutt, M. S. R., Wogan, G. N., and Davidson, C. S.,** Association between aflatoxin content of food and hepatoma frequency in Uganda, *Cancer,* 28, 253, 1971.
186. **Shank, R. C., Gordon, J. E., Wogan, G. N., Nondasuta, A., and Subhamani, B.,** Dietary aflatoxins and human liver cancer. III. Field survey of rural Thai families for ingested aflatoxins, *Food Cosmet. Toxicol.,* 10, 71, 1972.
187. **Peers, F. G. and Linsell, C. A.,** Dietary aflatoxins and liver cancer — a population based study in Kenya, *Br. J. Cancer,* 27, 473, 1973.
188. **Peers, F. G., Gilman, G. A., and Linsell, C. A.,** Dietary aflatoxins and human liver cancer. A study in Swaziland, *Int. J. Cancer,* 17, 167, 1976.
189. **Van Rensburg, S. J., Van Der Watt, J. J., Purchase, I. F. H., Pereira, Countinho, L., and Markham, R.,** Primary liver cancer rate and aflatoxin intake in a high cancer area, *S. Afr. Med. J.,* 48, 2508, 1974.
190. **Peers, F. G. and Linsell, C. A.,** Dietary aflatoxins and human primary liver cancer, *Ann. Nutr. Aliment.,* 31, 1005, 1977.
191. **Sun, T. and Wang, N.,** Studies on human liver carcinogenesis, in *Human Carcinogenesis,* Harris, C. C. and Autrup, H. N., Eds., Academic Press, New York, 1983, 757.
192. **Deger, G. E.,** Aflatoxin — human colon carcinogenesis? *Ann. Int. Med.,* 85, 204, 1976.

Chapter 2

AFLATOXIN AND REYE'S SYNDROME

I. INTRODUCTION

Aflatoxin has been considered one of the possible factors in the etiology of Reye's syndrome, the nature of which remains obscure although 25 years have passed since the first report on it. This mysterious children's disease, which attracts attention of clinical and laboratory researchers all over the world, is unquestionably a complex disorder, multifactorial in nature, and multisystem in involvement. This chapter reviews what knowledge has been acquired of this fascinating disease for a better understanding of its complexity.

II. HISTORY OF REYE'S SYNDROME

Reye's syndrome (RS) is a virus-associated biphasic disease that causes acute encephalopathy with fatty degeneration of the viscera almost exclusively in infants and children; it is very rare in adults.[1-3] The concurrence of acute brain swelling and fatty liver in the encephalitis-like illness of children was first noted by Brain et al.[4] in 1929 in six children 3 to 10 years of age and reported as "acute meningoencephalomyelitis of childhood". Although similar cases have been reported since that time under various titles, such as "Chickenpox associated with fulminating hepatitis",[5] "Fatal hypoglycemia in early non-icteric infectious hepatitis",[6] and "Acute encephalopathies of obscure origin in infants and children",[7] an interest in these encephalopathies of obscure origin was rekindled in 1963 by the Australian pathologist Douglas Reye[8] when he clearly defined the clinical and pathological features of this unique childhood syndrome which now bears his name. Retrospective surveys of autopsy records suggest this disease entity was reported only infrequently prior to the 1950s. The steadily increasing number of case reports throughout the 1960s and 1970s has proved that this syndrome is both widespread around the world and more prevalent than originally suspected. An intensive study, particularly of workers in the U.S. and especially of those at the Centers for Disease Control (CDC) in Atlanta, GA, has shed some light on most of the important questions concerning the disease, but its etiology remains still unknown.

A. CLINICAL SYMPTOMS

The clinical course of RS is usually biphasic and quite uniform, especially in children aged over 1 year. Previously healthy children contract a viral infection, most commonly influenza B[9] or varicella.[10] The process of recovery from this infection usually begins when the children develop repetitive vomiting, followed in 24 to 48 h by symptoms of CNS dysfunction, including lethargy, disorientation, and combativeness. These symptoms may lead rapidly to a coma with rostral-caudal progression to a brain stem dysfunction and death.[11] Respiratory disturbances, primarily hyperventilation, are frequent. The liver was enlarged slightly early in the illness in 30% of cases, but children were not jaundiced.[12] Seizures may occur at any time in the course of the disease. Evidence of increased intracranial pressure is present in the most severely ill patients.

B. LABORATORY DATA

Hematologic examination — The leukocyte count can range between normal and 20 to 40 $\times 10^9$/l, though the hemoglobin level and the erythrocyte and platelet counts are usually within normal limits.[13]

Serological findings — Serum transaminases (ALT and AST) are generally elevated. Their levels range from slightly above normal to several thousands units.[14,15] Elevated serum creatinine phosphokinase (CPK) and lactate dehydrogenase (LDH) have often been reported.[16,17] The presence of hyperammonemia clinches the diagnosis of RS. The blood ammonia level seems to peak in early phases of the disease, then it rapidly declines to normal values within 40 to 72 h.[18-22] An initial blood ammonia level of more then 300mg/dl has been associated with a significant increase in mortality.[23,24] However, other authors reported total recovery in patients who had blood ammonia levels well above 300 mg/dl.[18,22,25,26] Hypoglycemia is found in about 40% of patients and most often in patients under the age of 4.[27] The blood urea nitrogen and serum creatinine levels are usually within normal limits or only mildly raised.[28,29] Only in few cases of RS which ended in acute renal failure, elevated levels of urea nitrogen and creatinine have been reported.[30-32] A mild transient increase in the total serum bilirubin has been reported, although in most cases the bilirubin level was normal, and this fact helps to differentiate RS from most cases of acute hepatic encephalopathy.[27] The prolonged prothrombin time in patients with RS is often associated with hypofibrinogenemia and decreased levels of liver-dependent coagulation factors (V, VII, IX, and X).[33] Other laboratory abnormalities include lactic acidemia,[34] hyperaminoacidemia, and organic acidemia.[35,36]

Urine examination — Increased levels of amino acids and organic acids are found in RS cases.[8,37,38] In some instances, abnormal metabolites such as pteridines and hydroxypteridines have been reported.[39,40]

Cerebrospinal fluid — The cerebrospinal fluid pressure may be elevated, but cell and protein values are normal. The CSF sugar is usually low and tends to parallel the low blood sugar when these determinations are performed simultaneously.[27,37]

When encephalopathy advances to the stage when the patient is delirious, hyperventilation appears. This may be a result of either alkalosis or mixed respiratory alkalosis and mild metabolic acidosis, which is believed to be a result of stimulation of the brain stem.[11] Electroencephalography can assist in setting the severity of encephalopathy.

In 1974, Lovejoy et al.[21] described five stages based on the data of 40 patients suffering from RS. These stages have been helpful, permitting objective evaluation of a patient's progress by predicting the outcome and efficacy of therapeutic modalities. Children whose disease is at the stage IV or V, or is rapidly progressing from stage I to stage III, have a poor prognosis. The severity of the coma also correlated with the possibility of severe neurological residual impairment. Corey et al.[41] have reported that out of the survivors who had reached the stage IV or V during their illness, 30% of them had a residual neurologic deficit.

C. THERAPY

The most important factor in successful management of RS is an early identification of the victims and institution of an aggressive therapy as early as possible in an effort to maintain the case at stage I. RS is associated with a complex variety of metabolic disturbances; both death and the neurologic sequelae in survivors are attributed to an insult to the central nervous system. Considering that the etiopathogenic factors of RS are not clear, a causal therapy does not exist and management is symptomatic; the therapy is focused on maintaining an adequate cerebral perfusion pressure and controlling and minimizing the associated metabolic dysfunctions. Details of management vary somewhat in different centers, but an early diagnosis, constant monitoring, and adequate supportive care are among the important common features.

An outline of the treatment protocol for RS used in the Medical Center in San Diego, CA, is presented:[42]

1. Temperature should be maintained below 37°C with a cooling blanket.
2. Nasotracheal intubation is performed.
3. Controlled hyperventilation via a mechanical respirator should be used to maintain carbon dioxide tension at 25 mmHg.
4. Hypertonic glucose (15 to 20% solution) should be given intravenously to keep blood glucose levels between 150 and 200 mg/dl.

5. Insulin is administered intravenously, 1 unit/10 mg of glucose every 4 h initially; the dose is increased according to blood glucose concentration. Alternatively, insulin can be given as a continuous intravenous infusion if albumin is added to the intravenous solution. Dextrostix should be used to check blood glucose immediately before and a $1/2$ h after insulin has been given. A quantitative serum glucose determination should be performed every 4 h.
6. Fluid and electrolyte balance should be carefully maintained. A bladder catheter and a nasogastric tube should be inserted so that accurate intake and output can be calculated.
7. Neomycin enemas may be given to help reduce serum ammonia levels.
8. Central venous and arterial pressures should be monitored.
9. A ventricular or subdural monitoring device should be inserted for continuous recording of the intracranial pressure. If the prothrombin time is prolonged, fresh-frozen plasma can be given or a partial exchange transfusion performed to correct it, so that the monitor can be inserted safely.
10. If the intracranial pressure is elevated or if the patient is fighting the respirator, pancuronium bromide (Pavulon) in a dose of 0.1 to 0.2 mg/kg of body weight should be given for muscle paralysis after a careful neurological assessment has been completed. Pavulon dosing can be discontinued briefly if neurological reassessment is required, but the intracranial pressure should be watched closely, as it may rise.
11. Mannitol, 0.25 mg/kg/dose, should be given intravenously if the intracranial pressure rises above 20 mmHg.
12. Serum osmolality should be measured every 4 h and kept below 320 mOsm/l.
13. A careful chest physiotherapy and suctioning should be performed every 2 h. The patient should not be turned, but positioned on his back and his head elevated to 30°.
14. If these measures fail to control intracranial pressure, a barbiturate coma should be instituted.

The use of a barbiturate coma to decrease the intracranial pressure can, however, be associated with a hazard, as barbiturates can produce significant hypotension and thus reduce the cerebral perfusion pressure. During barbiturate therapy, the clinician loses the ability to assess the patient neurologically.[42]

Other therapeutic modalities have been described. Craniectomy has been used successfully in some cases of RS, and the clinical response of patients treated with craniectomy offers convincing evidence that encephalopathy is reversible if fatal consequences of brain swelling can be prevented.[11] Other authors have reported success in a therapy, such as the exchange transfusion providing "toxin" removal.[11,43,44] Another rationale for exchange transfusion was the need to correct clotting abnormalities. This method was not shared by other authors.[22,41,45] A total hypothermic body washout was proposed by Lansky et al.,[46] but little justification has been found for this technique. Samaha et al.[15] advocated peritoneal dialysis, but subsequently indicated that it was of little value. Charcoal hemoperfusion in children with RS has been used by Engle et al.,[24] based on the reported benefit of hemoperfusion in patients with fulminant hepatic failure. The protocol was terminated after the first three patients undergoing perfusion developed severe coagulation disturbances.[24,47]

It is clear that no totally satisfactory method is available for the treatment of RS. Nevertheless, at present, a multifaceted and aggressive approach to the treatment of the most serious symptoms — namely a marked elevation of the intracranial pressure and severe metabolic abnormalities — gives these children a favorable outcome and can reduce mortality.

D. MORPHOLOGICAL FEATURES
1. Gross Pathological Changes

Cerebral edema, usually grossly evident, is manifested as a heavy brain with swollen and flattened gyri and narrowed sulci. The liver is usually slightly enlarged and the capsular and cut surface are uniformly bright yellow. The kidneys have a mildly widened pale cortex with a faint yellow tinge. The heart is usually dilated, and sometimes the myocardium has a faint yellow color. In some cases the stomach contains a small quantity of dark fluid colored with blood, and occasionally multiple superficial erosions are present.[28]

2. Histological Changes

The cerebral changes include congested blood vessels, perivascular hemorrhages, and diffuse pericellular and perivascular edema, particularly in the cortex. There are anoxic neuronal changes consisting of shrunken, homogenous, eosinophilic cytoplasm and of pyknotic nuclei. Swollen astrocytes are also present.

Fatty metamorphosis of the liver is the most constant pathologic feature. The hepatic parenchymal cells are filled with numerous vacuoles which do not displace the nuclei and stain with Sudan. This fatty change is diffuse and uniformly distributed throughout the liver. In addition to the fatty change, glycogen in the liver cells seems substantially reduced or absent. Although cases with focal hepatic necrosis and slight mononuclear cell inflammatory infiltration in the portal tracts have been recorded,[40,48-51] they are unusual, and Bove et al.[52] considered it to be superimposed changes rather than features of this syndrome.

In the kidneys, fatty degeneration is obvious in the proximal convoluted tubules and in the loops of Henle. The glomeruli and interstitial tissue are entirely normal. In the heart, a fine droplet steatosis is present with a maximum of the subendocardial and subepicardial areas and in the bundle of His. An acute diffuse interstitial pancreatitis has been reported in some cases.[53,54]

3. Ultrastructural Changes

The ultrastructural changes in the liver, brain, and skeletal muscles of patients with RS have been first described by Partin et al.[55-57] It has been found that the changes are different at various stages of the disease.

In the earliest stage of hepatic injury, mitochondria exhibit mild matrix expansion and pleomorphismus. The degree of mitochondrial involvement varies from cell to cell. There is a progressive increase in smooth endoplasmic reticulum. Peroxisomes first increase in number and, in more severely affected cases, exhibit a watery alteration of their matrix. Within the smooth endoplasmic reticulum, there is abundant triglyceride in small nonmembrane-bounded cytoplasmic droplets. Nuclear changes consist of a loosening of the nucleolar structure and of a burst of mitotic activity.

The ultrastructural changes in the liver biopsy specimens have been summarized by Partin[57] as follows:

1. Mitochondrial changes: matrix expansion, progressive loss of matrix dense bodies, matrix disorganization, gross swelling, and outer membrane rupture
2. Cytoplasmic changes: glycogen depletion, smooth endoplasmic reticulum increase, triglyceride accumulation in endoplasmic reticulum, Peroxisome increase — alteration, decreasing Golgi VLDL content, watery expansion of cytosol, increased lysosomes, and increasing cell volume
3. Canalicular changes: progressive dilatation, microvillus alteration, and luminal accumulations
4. Nuclear changes: nuclear "skein"-ing, mitotic burst during early recovery, and nuclear swelling and disorganization of chromatin in fatal cases

The cardinal ultrastructural changes in the brain are astrocyte swelling and partial deglycogenation, myelin bleb formation, and universal injury to neuron mitochondria. The mitochondrial injury consists of matrix disruption with a moderate but not massive matrix swelling. Dilatation of the rough endoplasmic reticulum and nuclear changes occur only in neurons with

TABLE 2
Metabolic Abnormalities in Reye's Syndrome[a]

Hyperammonemia	Hypocholesterolemia
Hypoglycemia	Hypophosphatemia
Hypertransaminasemia	Hypocomplementemia
Prolonged prothrombin time	Increased metabolic rate
Hyperaminoacidemia	Dicarboxylic aciduria
Hyperuricemia	Hypocetonemia
Lactic acidosis	Low mitochondrial enzymes
Free fatty acidemia	Increased cortisol
Short chain fatty acidemia	Decreased hepatic glycogen
Organic acidemia (i.e., propionic)	Elevated CPK
Hypercatecholaminemia	Fatty liver
Respiratory alkalosis	Abnormal mitochondria

[a] From Glasgow.[58]

severely altered mitochondria. The mitochondrial ultrastructure of cerebral neurons resembles the unique mitochondrial ultrastructural changes seen in the liver parenchyma in RS.[56]

Ultrastructural changes of the skeletal muscles have two types of mitochondrial alteration. The first, affecting the majority of mitochondria, consists of a moderate matrix swelling which produces mitochondrial profiles that are too smooth and that present slight to moderate alterations of matrix density. The second type of injury consists of frank disruption of the mitochondrial matrix. This kind of injury is associated with frank myolysis and disorganization of myofibrils and myofilaments. Unlike the process in hepatocytes, in which all mitochondria seem equally susceptible to injury, the muscle cell seems to possess a major population of mitochondria which are relatively resistant to matrix alteration and a minor population of those which are quite susceptible to it.[57]

E. PATHOGENESIS

Although a lot of new information involving the pathogenic mechanism of RS has been reported since Reye's article[8] appeared, the basic pathological process remains unclear. Mitochondrial injury has been dominating theory on the pathogenesis of this disease since Partin et al.[56,57] published their findings on alterations of mitochondrial ultrastructure in the liver, brain, and skeletal muscle cells. These authors suggest that the mitochondrion is the principal site of injury in RS, and this organelle pathology appears to be highly characteristic of the disease. The mitochondrial lesions parallel the severity of encephalopathy. In addition to the morphological evidence of mitochondrial injury, there is a biochemical evidence implicating mitochondrial dysfunction. The list of biochemical abnormalities in RS is very long, as indicated in Table 2, reported by Glasgow.[58]

Many, if not all, of these metabolic disorders seem to be explainable by mitochondrial dysfunction.

On the clinical level, three functional abnormalities are consistently associated with RS: prolonged prothrombin time, elevated serum transaminases, and hyperammonemia.

Prolongation of the prothrombin time is usually associated with severe liver damage, and therefore, this finding is unusual in RS with its well-preserved liver parenchyma. It may reflect selective inhibition of the clotting factor synthesis or release.[59]

The serum transaminases in RS are elevated and often change with remarkable rapidity. Not infrequently, extremely high transaminase levels are usually found only in the liver, accompanied with extensive necroses or inflammatory processes.

Therefore, the sources of these strikingly rapid elevations in transaminases in RS are difficult to explain. Thaler[59] demonstrated a reversal in the cytoplasmic-to-mitochondrial SGOT ratio in

RS patients and proposed a hypothesis that dislocation of protein such as SGOT from the mitochondrial matrix to the cytoplasm may increase the permeability of the mitochondrial membranes, and thus losses of other osmotically more active molecules from mitochondria may explain swelling of these organelles. Less likely is the possibility that an increase in mitochondrial SGOT reflects a prolongation of the relatively brief turnover rate of this isoenzyme compared with its cytoplasmic analog.

Elevated blood ammonia, first described by Huttenlocher[60] in 1969, is consistently present during the acute phase of this syndrome. A number of investigators have shown that the level of blood ammonia in RS correlate well with the course, severity, and stages of the disorder and also with its mortality rate.[21,41,61] However, a controversy still persists on this, because a few patients with RS did not develop hyperammonemia, and in some patients, the levels of serum ammonia did not correlate with the severity of the illness.[62,63] This biochemical abnormality suggests a metabolic block in urea synthesis. Several inherited defects in ammonia metabolism are known, the most common being the ornithine transcarbamylase (OTC) deficiency. Thaler and collaborators[64] originally described a patient with RS whose ornithine transcarbamylase (OTC) activity was depressed and concluded that RS was associated with a specific error of urea-cycle metabolism. In addition to this patient, they subsequently observed two boys with RS whose OTC level was depressed during the acute phase and remained depressed following a complete recovery from the illness.[59] They have postulated that a patient with this defect would be relatively protein tolerant, in contrast with the classical X-linked OTC deficiency, but vulnerable to illnesses associated with anorexia and vomiting, where a reduced intake of ornithine could limit the capacity of the urea cycle for a removal of ammonia.

Brown et al.[65] have published data on four patients with RS, indicating the reduced activity of OTC and carbamyl phosphate synthetase (CPS) during the first days of the clinical symptoms, which returned toward normal values during the following week, regardless of whether the disease ended in death or recovery. The authors therefore suggested that these metabolic abnormalities are acquired and transient.

Hyperaminoacidemia has been reported in RS cases.[66] The most striking and consistent elevations of serum concentrations occurred in glutamine, alanine, α-amino-N-butyrate, and lysine. Since Kang et al.[67] described two patients with a deficiency of OTC, and as one had a pattern of hyperaminoacidemia resembling the pattern observed with RS, it has been postulated that the amino acid pattern of RS may reflect the same metabolic disorder — a deficiency of OTC.

Trauner et al.[62] reported a 3- to 20-fold increase in the concentrations of short-chain fatty acids (propionate, butyrate, isobutyrate, valerate, and isovalerate) in the serum of patients in the acute phase of RS. Other investigators have reported elevated levels of the medium- and long-chain free fatty acids in sera of patients with RS during the acute illness, which returned to normal after recovery.[68,69] Ogburn[69] showed that the polyunsaturated fatty acid content of free fatty acids was increased in sera of patients with RS during the illness and suggested that since the polyunsaturated fatty acids are precursors of prostaglandins, the pattern of prostaglandins may be disturbed. The cause of free fatty acidemia and excessive lipolysis in RS is unclear. Trauner[70,71] speculated that one possible mechanism could be a block in the β-oxidation of fatty acids, e.g., either by the virus alone or in combination with an exogenous toxin.

Direct evidence for this is more circumstancial and includes accumulating substrates (i.e., fatty infiltration and free fatty acidemia), increased levels of side reaction metabolites (i.e., adipic, suberic, and subacid acids), and decreased product (i.e., hypoketonemia).

Roe et al.[72] observed dicarboxylic aciduria in RS patients, which signifies defective mitochondrial β-oxidation. Since a decreased level of plasma carnitine, an important enzyme involved in the transport of fatty acids into the mitochondria and necessary for β-oxidation, was noted in patients with RS, they have suggested that a metabolic injury may involve these pathways.

It has been shown on an animal model that short-chain fatty acids are hepatotoxins and neurotoxins causing neurological impairment and cerebral edema, and therefore, some investigators believe that free fatty acids may play an important role in the pathogenesis of RS.[70,71,73]

Hypoglycemia, which occurs in some patients, namely in infants with RS, can be also explained by a mitochondrial defect.[58] Hypoglycemia is usually readily treated with a relatively small amount of glucose suggesting that it is not a result of increased glucose consumption. Furthermore, serum insulin levels were appropriate. Liver glycogen is depleted in patients with RS who are hypoglycemic, and there is no glycemic response to injections of glucagon. This may be so due to glycogen depletion and not to the impairment of glycolysis, which is a cytoplasmic pathway. A defect in gluconeogenesis in RS has been suggested.[74,75] This presumption is supported by evidence of decreased activities of pyruvate dehydrogenase and pyruvate carboxylase, as well as by *in vitro* studies of RS liver showing a defect in the glucose formation from pyruvate.

Hypercatecholaminemia is a common metabolic feature in RS. A significant elevation of tyramine and octopamine concentrations in the plasma, urine, and brain of the encephalopathic patients with RS have been reported.[76,77] Decreased activity of monoamine oxidase, a marker enzyme for the outer mitochondrial membrane, has been considered the cause of an abnormal accumulation of these biogenic amines.[78] A central release of catecholamines may be involved in the development of cerebral edema and a specific relationship may exist between the increased concentration of plasma catecholamines and intracranial hypertension.[79]

Biochemical studies are consistent with ultrastructural mitochondrial abnormalities, namely there is a universal decrease in the activity of all mitochondrial enzymes examined to date. On the contrary, the investigated cytosolic enzymes exhibited normal activities.[80] Mitchell et al.[78] proposed a hypothesis that RS may reflect a generalized decrease in the steady-state level of mitochondrial enzymes due to a selective impairment of mitochondrial biogenesis which may be the primary lesion.

This controversial aspect involves the following question: Is encephalopathy secondary to metabolic abnormalities or is encephalopathy a direct result of the same basic pathological process?

Partin et al.[56] found that changes in the brain mitochondria are similar to those found in the liver and therefore suggested that both the organs are attacked simultaneously by the same toxic agent. However, Robinson et al.[81] showed that the brain mitochondria changes are not accompanied with a decrease in mitochondrial enzyme activity as seen in the liver and proposed that mitochondrial damage seen in the brain and muscle cells of patients with RS is either a secondary phenomenon resulting from a metabolic disturbance, or a representative of the same primary process that occurs in the liver mitochondria, but without the same biochemical sequelae. Other authors hold the view that encephalopathy may be secondary to hepatic mitochondrial injury.[34,60,62,76]

On the whole, available evidence suggests that encephalopathy is at least in part secondary to hepatic metabolic dysfunctions. In fact, at present there is very little information on what we could differentiate between these two possibilities.

F. EPIDEMIOLOGICAL STUDY IN THE U.S.

Most of the credit for the knowledge about RS is given to researchers in the U.S., particularly those at the Center for Disease Control (CDC) for their continuous epidemiological studies since 1973.[82,83] The CDC has maintained an interest in the epidemiology of RS since it investigated an outbreak of unusual encephalopathy in North Carolina in 1963, which in retrospect was probably the first outbreak of RS described in the U.S.[84]

The CDC first initiated nationwide surveillance on RS in 1973—74 through May 1984 with the exception of 1975.[82,85] Since 1973, more than 2800 cases of RS have been reported to CDC through the National Reye's Syndrome Surveillance System. For epidemiologic purposes, the CDC has defined RS cases as:[86]

1. An acute noninflammatory encephalopathy documented by

   a. Clinical picture, stage I or greater and, if available,
   b. CSF (cerebrospinal fluid) containing less than or equal to 8 WBC/mm,^{3} or
   c. Histological sections of the brain demonstrating cerebral edema without a perivascular or meningeal inflammation.

2. Evidence of hepatic involvement documented by either

   a. Microvesicular fatty metamorphosis of the liver, diagnosed by autopsy or biopsy, or
   b. Threefold or greater rise in the serum SGOT, SGPT, or NH_3, or
   c. No other more reasonable explanation for the cerebral and hepatic abnormalities.

1. Antecedent Illnesses and Seasonal Distribution of RS

The CDC has recently separated cases of RS into five groups according to the history of antecedent illnesses[86]:

1. Upper respiratory illnesses
2. Varicella infection
3. Gastrointestinal illness
4. Fever or nonvaricella rash or other unspecified illnesses
5. No history of an antecedent illness

The first three types of antecedent illnesses are most frequently associated with RS. Data collected through nationwide surveillance during the 5-year period (1974 and 1977 to 1980) have shown that 60 to 70% of RS cases have been associated with respiratory infection, 20 to 30% with varicella, and only 5 to 15% with antecedent gastrointestinal illnesses.[9]

The highest incidence of RS in the U.S. during the 5-year surveillance period was observed during periods of the major influenza B and influenza A (H_1N_1) activity and was somewhat lower during the single period of the influenza A (H_3N_2) epidemic. In the years 1981 to 1983, when influenza A (H_3N_2) was predominant, the incidence of RS was similar to that observed in 1978. In 1982, influenza B was the predominant influenza isolate, but the influenza activity in that year was relatively mild, and the reported incidence of RS was low.[86]

Reported incidence and case fatality ratio of RS in the U.S., 1973 to 74 and 1976 to 84 are indicated in Table 3.

Cases of RS occurred in all the months of the year; however, during the 5-year surveillance, they peaked during the influenza season, December through March.[9] In addition to influenza B and A, other viral agents such as adenoviruses, parainfluenza, herpes simplex, and cytomegalovirus associated with an antecedent upper respiratory infection were reported.[82] No distinct seasonal occurrence was observed in patients with gastrointestinal symptoms.

2. Age

RS is a disease of childhood. More than 90% of the total of cases reported each year were persons less than 15 years of age.[86] Patients with antecedent respiratory illnesses (mean age between 8 and 11 years) were more numerous than those with varicella or gastrointestinal prodromes (mean age of 6 years). These differences could reflect the age-specific attack rates of these antecedent viral illnesses.[9,85]

3. Sex and Race

RS cases were nearly equally divided between males and females, except for 1976 to 1977 with a slight predominance of female patients.[9] Nationwide, 90% or more of the total of cases

TABLE 3
Reported Incidence and Case Fatality Ratio of Reye's Syndrome in the U.S., 1973—1974 and 1976—1984[a]

Major year	Influenza activity	No. of cases	Incidence	Death case ratio	
				No. of cases	Death ratio (%)
1974	B	379	0.58	157/379	41
1977	B	454	0.71	156/373	42
1978	A/H$_3$/N$_2$	237	0.37	66/225	29
1979	A/H$_1$N$_1$	389	0.62	113/349	32
1980	B	555	0.88	114/516	22
1981	A/H$_3$N$_2$/H$_1$N$_1$	297	0.47	83/288	29
1982	B	213	0.34	70/210	33
1983	A/H$_3$N$_2$	198	0.32	52/191	30
1984	A/H$_1$N$_1$/B	75		14/71	20

[a] From Barrett et al.[86]

reported each year were in whites. The age distribution of black and white children has been found different. In white children, RS cases were reported in all age groups, but only a small percentage of them were children under 1 year of age, whereas in blacks, a large proportion of cases reported were infants under 1 year of age.[86]

4. Mortality

Since 1974, the case fatality rate of RS determined through nationwide surveillance has declined from 42% in 1974[83] to 30% in the years 1978 to 1983.[86] The case fatality rates were higher among respiratory than varicella-associated RS cases. Prognosis for infants under 1 year of age is poor.[83] The decreased mortality in recent years is probably due to increased reporting of nonfatal cases and the recognition of earlier and milder cases of the syndrome. There is evidence that earlier treatments and improvements in therapy have contributed to this decrease.[44,85]

5. Rural Clustering

In those less than 18 years of age, an incidence rate of 1.80 cases/100,000 of population has been recorded in rural areas compared with 0.42/100,000 in urban and 0.58/100,000 in suburban areas.[82]

G. EPIDEMIOLOGICAL STUDIES OUTSIDE THE U.S.
1. Reye's Syndrome in Great Britain

The British Reye's Syndrome Surveillance Scheme (BRSSS) was set up in August 1981. Results of this surveillance period (1981 to 1983) have shown some differences compared with the data from the U.S.[87] Seasonal distribution of the marked winter peak, repeatedly demonstrated by the CDC studies, has not been found. There was an excess of cases in the autumn and winter months compared to the spring and summer ones. The marked association of RS cases with influenza B and A has not been demonstrated. In 15 out of 57 patients, laboratory-confirmed viral infection cases, only three patients had laboratory-confirmed influenza (two type A, one type B).[87] The mean and median ages were less than half those reported by CDC, with a high proportion of cases under 1 year of age. No case has been reported in black infants. Two epidemiological features were similar to the North American findings — a slight excess of female cases and a striking excess of cases in rural areas.[87]

TABLE 4
Viruses Associated with Reye's Syndrome

Virus	Ref.
Varicella	5, 8, 10, 29, 44, 50, 60, 89—98
Influenza B	10, 99—109
Influenza A	85, 96, 105, 110—117
Coxsackie B_1, B_2, B_4, A	84, 90, 118—121
Adenovirus 3, 7	96, 109, 118, 122—126
Echo	84, 106, 127
Herpes Simplex	121, 127—129
EPB (Epstein-Barr virus)	53, 130, 131
Morbilli	92, 127
Parainfluenza	96, 132
Reovirus	121, 133
RS virus	109, 134
Myxovirus	33
Polio type 1	135
Rubeola	136
Live virus vaccine	137

2. Reye's Syndrome in Asia

Epidemiological studies on RS in Asia based on nationwide surveillance in Japan (1978 and 1983) and on research studies carried out in Korea, Taiwan, Thailand, Malaysia, Singapore, and Indonesia have shown some differences compared with the results obtained in the U.S.[88] In Asian countries, upper respiratory infection and gastrointestinal disorders were the dominant antecedent illnesses. However, 25% of patients in Japan and 11% in Korea had no antecedent illnesses. The incidence of preceding varicella was very low. There was a relatively high rate of preceding measles and dengue fever in Indonesia. In 1978, winter and summer were the peak seasons in Japan. The second survey (1982) showed the highest incidence in winter during a nationwide epidemic of influenza B. The age peak incidence varied between the first, second, and third years of age. Case fatality rate remained higher in Asian countries.

H. ETIOLOGY

Four concepts have been implicated in the etiology of RS:

1. Viral infection
2. Genetic factors
3. Intrinsic and extrinsic toxins
4. Interaction of viruses and toxins

1. Viral Infection

It is well established that RS is preceded by a viral illness, usually an upper respiratory tract infection, due to influenza or varicella. Epidemiological studies in the U.S. have shown that a large number of cases were associated with influenza B infection.[82,86,98] Most cases of RS peaked during the winter and spring months when respiratory illnesses are epidemic.[86,98] The cases associated with varicella are sporadically distributed throughout the year.[86] Numerous other viruses associated with RS have been, however, reported in literature as indicated in Table 4.

Other studies have shown that a mixed viral infection consisting of herpes virus and a myxo- or paramyxovirus is also implicated in this disease.[96,105,138] This great variety of viruses indicates that RS is not a sequel of specific infection, but it is assumed that the disease is "triggered" by a multiplicity of viral agents.[132]

TABLE 5
**Reye's Syndrome Cases with Positive Isolation
of Viruses**

Virus	Ref.
Influenza B	10, 96, 100, 101, 105, 107
Influenza A	85, 117, 137, 139
Adenovirus 3, 7	96, 123, 124, 126
Coxsackie A	90, 96, 121
Coxsackie B	90, 119, 120
Reovirus	120, 121
Myxovirus	33
Herpes Simplex	121, 127, 129
Echo	101, 127
Parainfluenza	132

It was also regarded unlikely that RS could be the direct effect of a virus because a virus has only rarely been isolated from altered tissue (see Table 5).

Furthermore, the absence of conventional signs of infection, inflammation, or necrosis in the liver and other involved organs has been considered evidence against a direct viral effect.

On the other hand, epidemiological studies in the U.S. have increasingly associated RS with two viral agents — influenza and varicella. Influenza virus was isolated in large amounts from the lung, liver, and heart, showing that influenza virus may be present in the affected tissues of RS patients without histological features of cell injury.[107,117] The hypothesis of a direct virus cause seems to be supported by reporting virus-like particles found in RS tissues by electron microscopy.[140-143] Davis et al.[144] observed on their mouse model of RS that intravenous infection with high titered pools of influenza B virus caused many of the clinical, biochemical, and pathologic features of the mouse illness similar to those seen in RS, although virus replication was not required for the effect. Since virus replication is not needed to induce changes, the possibility of organ localization of circulating viral antigen as a cause of cellular abnormality has been considered. Toxic effects induced by viral antigen were provided by a report of circulating adenovirus penton antigen, known to be cytotoxic *in vitro* in three children with adenovirus-type-7 pneumonia and significant extrapulmonary diseases simulating RS.[124] It has also been suggested that metabolic abnormalities may be related to a direct effect of viruses. Younkin and Gudzimovicz[145] assume that the activation of ornithine decarboxylase by influenza viral RNA polymerase may be an initiating factor followed by metabolic abnormalities and mitochondrial injury in RS.

An alternative possibility of a direct viral effect is presented by inducing an immune response that alters the normal cellular metabolism and thus initiates metabolic abnormalities. Immunoglobulin deposits in the vessels of skeletal muscles in RS cases, also exhibiting the characteristic mitochondrial changes, have been reported.[146] Whereas Linnemann et al.[105] in his study was not able to find an impairment of cellular immunity as reflected in the lymphocyte transformation by phytohemagglutinin or influenza B antigen, Tang et al.[147] have recently described tubuloreticular inclusions in blood lymphocytes and monocytes in two cases of RS initiated by influenza infection during the acute phase of the illness. Since these inclusions were demonstrated in patients with acquired immunodeficiency syndrome, it has been postulated that the finding of these inclusions in RS may reflect a viral infection and/or an immune dysfunction.

2. Genetic Factors

In some RS cases, genetic abnormalities or defects were reported.[59,64,148] It has been therefore suggested that genetic factors may affect a child's susceptibility to RS following viral infection. This suggestion has been supported by reporting cases of RS in twins[149] and siblings.[19,110,150]

Thaler and collaborators[64] originally described a patient with RS, whose OTC level was depressed during the acute phase of the illness and remained depressed following his complete recovery, but Brown et al.[151] have demonstrated that the changes in levels of OTC seen in their RS cases are an acquired and transient dysfunction of the urea cycle enzyme. While there is no doubt that the urea cycle enzymes are depressed during the acute phase of the illness, the child originally described by Thaler[59] probably had an inherited deficiency of OTC with symptoms similar to RS. More recently, two other patients with partial OTC deficiency simulating RS have been reported by Yokoi et al.[152]

Initially, it was felt that high mortality precluded recurrences, but in recent years, the apparent mortality rate has declined and more patients with "recurrent Reye's syndrome", having the classical prodromes and course of RS, have been recognized as having inherited metabolic diseases. Systemic carnitine deficiency[72,153] medium- and long-chain acyl-CoA dehydrogenase deficiency[154,155] have a close similarity to RS and may be easily confused with RS. However, there remains a much larger group of children with RS in whom no genetic error has been demonstrated.

3. Intrinsic and Extrinsic Toxins

Injury to the mitochondria described in RS suggested to many investigators the possibility that a toxic substance may play a role in the pathogenesis of this disease. The idea that there could be a single causative factor related to a viral prodrome initiating mitochondrial injury led to testing RS serum for endogenous toxic activity affecting bioenergetic functions of the isolated animal mitochondria. Aprille[156] has demonstrated a substance of low molecular weight in the serum of patients with RS that causes alterations of the mitochondrial structure and metabolism similar to those in RS. Although this factor was first thought to be unique in RS, it has subsequently been found in other circumstances, such as Chédiak-Higashi syndrome, chickenpox, hepatitis stage III, liver failure, and a sudden coma of unknown cause.[157] The main component of this intrinsic toxin has recently been identified as uric acid.[158] Probably the sick child retains the side products of cell metabolism, which is one way mitochondrial dysfunction can occur in RS.

In the initial description of this syndrome, Reye et al.[8] compared the disease to Jamaican vomiting sickness, described by Scott[159] in 1917. This later disease is thought to be caused by ingestion of the ackee fruit, which contains a toxic substance, hypoglycin. An experimental model has been developed in rats using 4-pentenoic acid, an analog of hypoglycin. Due to biochemical and morphological similarities to the features of RS, Jamaican vomiting disease and pentenoic acid have been suggested to relate to or perhaps share common pathophysiology.[160]

A number of other chemicals and drugs which induce hepatic and central nervous system (CNS) disorder similar to RS, such as valproic acid,[161,162] methyl bromide,[163] pyrrolozidine (Senecio),[164] margosa oil,[165] camphor,[166] and insect repellent,[167] have been reported.

Most attention regarding the toxin hypothesis has been particularly directed toward the possible role of salicylates, environmental chemicals, and aflatoxins.

4. Salicylates

An association between RS and salicylates has been suspected for a long time. Mortimer et al.[168] and Giles[169] were the first who noted similarities between the encephalopathy of RS and the symptoms associated with salicylate toxicity. A hepatic dysfunction and encephalopathy occur in both cases and biochemical abnormalities including hypoglycemia, coagulopathy, elevated SGOT values, and mixed acid base disturbances are associated with both.[170,171] However, there exist important differences, including the nature of amino acidemia,[36] the lack of abnormalities in the urea cycle enzymes,[172] and liver ultrastructural changes.[173-175]

Epidemiological studies performed during the last 5 years have suggested a definite association between RS and ingestion of salicylates during the antecedent illness.[176-178] However, several investigators have raised doubts about the validity of these results because of possible biases in the epidemiological studies, such as selection, recall, and misclassification.[172,179-181] The serum salicylate concentrations reported in patients with RS often occur within the therapeutic range. However, these salicylate concentrations may be inappropriately high when given the time interval between the last dose of aspirin and measurement of their serum concentration. Partin et al.[182] reported significantly higher salicylate concentrations in RS patients who died or had serious neurologic deficits when compared to patients who survived without neurologic sequelae. Because salicylates are known as mitochondrial toxins and the mitochondria are known to be significantly injured in RS, Partin suggested that the use of aspirin in children during RS outbreaks or in children suffering from influenza and varicella should be avoided.

Because of many differences between salicylate toxicity and RS, it seems unlikely that salicylates alone could be the cause of RS. Three hypotheses have been proposed for the possible role of salicylates. First, salicylates may be acting on certain viruses only as an additive or synergistic toxin, leading to an accentuated degree of mitochondrial damage.[177] Second, Bailey et al.[183] have suggested that salicylates may alter the immune system, which leads to increased viral virulence. Third, there is the hypothesis that a genetic defect in salicylate metabolism may be present in RS patients.[184]

Although the present data strongly suggest a role played by salicylates in conjunction with the virus, in the etiology of RS, other factors, such as genetic or environmental, may be involved in the etiology of RS.

5. Environmental Toxins

The predominantly rural distribution of RS, the clustering of cases in certain regions, and the inexplicability of such clustering based on the incidence of viral diseases in the population led investigators to the question of whether environmental toxins such as herbicides, insecticides, and surfactants can alter a child's response to viral infection.[185-187] To test this hypothesis, several animal models have been developed. Crocker et al.[185] painted newborn mice with pesticides and/or solvents and then injected them with a sublethal dose of mouse encephalomyocarditis virus. They found in these animals an increased mortality rate and morphological changes similar to those in RS. Further studies of this group revealed that insecticides and emulsifiers were also able to enhance the effect of the virus.[188] Colon and associates[189] pretreated juvenile rats with various chemical toxins and then injected them with an ineffective dose of the Mengo strain of encephalomyocarditis virus. They reported that of the chemicals tested, pentenoic acid showed a maximum synergism on viral infection in producing symptoms of RS. Hug et al.[190] studied the effects of several chemicals such as salicylates, fructose, atlox, sodium pentachlorophenol, and butyrated hydroxytoluene, with and without EMG virus infection in weanling mice. They showed that the treatment of mice with the virus and atlox or with butyrated hydroxytoluene induced not only morphological changes in the liver, similar to those in RS, but also biochemical abnormalities such as decreased activity of CPS and OTC. These experimental results indicate that there is an apparent association between the virus and certain chemicals, resulting in increased morbidity and mortality.

6. Aflatoxins

The possibility that some cases of RS might be associated with aflatoxin ingestion was first suggested by Becroft,[129] but identification of this toxin was first reported in Thai children with Udorn encephalopathy, an illness that resembles RS.[50,191] Forty cases of encephalopathy and fatty degeneration of the viscera in Thai children, confirmed at the autopsy or biopsy, were

TABLE 6
References of Reye's Syndrome Cases Associated with Aflatoxin

1.	Bourgeois, C. H., et al.[48]	1971	Thailand
2.	Olson, L. C., et al.[191]	1971	Thailand
3.	Shank, R. C., et al.[194]	1971	Thailand
4.	Becroft, D. M. O.[195]	1972	New Zealand
5.	Dvořáčková, I., et al.[196—198]	1972	Czechoslovakia
		1974	
		1977	
6.	Chaves-Carballo, E., et al.[199]	1976	U.S.
7.	Hogan, G. R., et al.[200]	1978	U.S.
8.	Ryan, N. J., et al.[201]	1979	U.S.
9.	Nelson, D. B., et al.[202]	1980	U.S.
10.	Wulur, H. (Ref. Yamashita, F.[88])	1981	Indonesia
11.	Rogan, W. J.[203]	1985	U.S.

characterized by an abrupt onset of coma or convulsions, fever, respiratory distress, vomiting, and death within 72 h. Abnormal laboratory findings included elevated serum transaminases and free fatty acids and prolonged prothrombin time. The spinal fluid was clear with a low glucose level. Autopsies revealed (1) cerebral edema and neuronal degeneration, (2) fatty degeneration of the liver, kidney, and heart, and (3) generalized lymphocytolysis. Chemical assays were performed on the brain, liver, kidney, stomach and intestinal content, and stool specimens from 22 out of 40 children. Aflatoxin B_1 was recovered in 96% of these samples.

Epidemiological data have shown a clustering of cases in Northeastern Thailand, which occurred mainly in the rural areas and were geographically and seasonally related to the high levels of aflatoxin contamination of market food samples.[191,192] In two cases, the presence of heavy aflatoxin contamination in leftover foods eaten 2 or 3 d prior to the onset of the disease was demonstrated.[50,192] Convincing evidence of the role of aflatoxin in this syndrome has come from experimental animal studies. A food sample collected from one patient's home yielded a number of toxinogenic strains of fungus, principally *Aspergillus flavus*. The purified toxin from this culture, when administered to juvenile macaques, reproduced the clinical, biochemical, and pathological features seen in Thai children.[193]

The documentation of hepatotoxicity of aflatoxin B_1 and the recovery of AFB_1 from the tissue of children in Thailand stimulated an interest in AFB_1 and its possible role in the etiology of RS. Several subsequent reports have then demonstrated aflatoxin in the tissue, blood, and urine of RS patients (Table 6).

These findings have led to the suggestion that aflatoxin, which has been recognized to contaminate a great quantity of food, including the infant formula,[204] may be involved in the etiology of RS.[193-205]

7. Interaction of Viruses and Toxins

Experimental data indicate that there is an apparent synergism of viruses and certain chemicals, resulting in increased morbidity and mortality in animals,[185,186,206] insects,[207] and cell cultures.[208,209] As this pathologic association can occur experimentally, it has been suggested that it may also occur in humans.

Pollack[210] speculated that under the stress induced by a virus or viruses, lipolysis of depot fat is associated with an abnormal lipid metabolism responsible for the release of high levels of toxic free fatty acids or other innocuously accumulated toxic chemicals in the adipose, which may lead to structural and biochemical abnormalities.

Mullen[211] evaluated the laboratory, clinical, and epidemiological data of RS from biochemical and immunopharmacological aspects and concluded that environmental exposure to various potentially toxic compounds, or their biotransformation products called xenobiotics, may be responsible for the early pathogenesis of RS. He proposed that the xenobiotic factor or factors may initiate biochemical and functional abnormalities of the liver with a subsequent impairment of normal immune defense against viral infection. The prodromal illness leads to further biochemical and immunological abnormalities, either exacerbated by the toxic residuum or still abnormal cellular function, progressing to fatty degeneration of the liver and encephalopathy.

The presumed impairment of immune defense in RS patients has been supported by recent reports.[147,212] Rozee et al.[212] demonstrated that lymphocytes obtained from patients during the acute phase of RS responded very poorly to exposure to Newcastle disease virus, and produced significantly less interferon than lymphocytes of children during the convalescence or those of the controls. Since interferon is one of the important factors in recovery from virus diseases, it has been suggested that a compromised interferon response may be an important factor in RS.

Tang et al.[147] found tubuloreticular inclusions in blood lymphocytes and monocytes of RS patients. Since these inclusions were found in monocytes of patients with acquired immunodeficiency syndrome, they assumed that this finding in RS might reflect an immune dysfunction.

Though all these above-mentioned etiological concepts are often discussed separately, it is probable that all of them contain an element of validity. Certainly, viral exposure is quite common in most cases and appears to be essential for the development of this syndrome. Why most children who manifest this unique response on one occasion fail to have similar episodes following subsequent viral infection remains unexplained, but it seems likely that extrinsic toxins act as primary factors in the genesis of this disease.[12]

REFERENCES

1. **Vanholder, R., De Renck, J., Sieben-Praet, M., and DeCoster, W.,** Reye's syndrome in adult, *Eur. Neurol.,* 18, 367, 1979.
2. **Varma, R. R., Riedel, D. R., Komorowski, R. A., Harrington, G. J., and Nowak, T. V.,** Reye's syndrome in nonpediatric age groups, *JAMA,* 242, 1373, 1979.
3. **Maass, V., Holz-Slomczyk, M., and Schmidt, F. W.,** Reye-Syndrome im Erwachsenenalter, *Dtsch. Med. Wochenschr.,* 103, 1431, 1978.
4. **Brain, W. R., Hunter, D., and Turnbull, H. M.,** Acute meningo-encephalitis of childhood, *Lancet,* 1, 221, 1929.
5. **Breen, G. E. and Edmond, R. T. D.,** Chickenpox associated with fulminating hepatitis, *Med. Press,* 232, 241, 1954.
6. **Tomlinson, B. E.,** Fatal hypoglycemia in early non-icteric infective hepatitis, *Lancet,* 1, 1300, 1955.
7. **Lyon, C., Dodge, P. R., and Adams, R. D.,** The acute encephalopathies of obscure origin in infants and children, *Brain,* 84, 680, 1961.
8. **Reye, R. D. K., Morgan, G., and Baral, J.,** Encephalopathy and fatty degeneration of the viscera: a disease entity in childhood, *Lancet,* 2, 749, 1963.
9. **Hurwitz, E. S., Nelson, D. B., Davis, C., Morens, D., and Schonberger, L. B.,** National surveillance for Reye's syndrome: a five year review, *Pediatrics,* 70, 895, 1982.
10. **Glick, T. H., Likosky, W. H., Levitt, L. P., Mellin, R., and Reynolds, D. W.,** Reye's syndrome: an epidemiologic approach, *Pediatrics,* 46, 371, 1970.
11. **Partin, J. C.,** Reye's syndrome (encephalopathy and fatty liver). Diagnosis and treatment, *Gastroenterology,* 69, 511, 1975.
12. **Crocker, J. F. S. and Bagnell, P. C.,** Reye's syndrome: a clinical review, *Can. Med. Assoc. J.,* 124, 375, 1981.
13. **Muller, P. and Tonz, O.,** Reye's syndrome: Enzaphalopathie und Fettdegeneration der inneren Organe, *Helv. Paediatr. Acta,* 26, 371, 1971.
14. **DeVito, D. C.,** How common is Reye's syndrome? *N. Engl. J. Med.,* 309, 179, 1983.

15. **Samaha, F. J., Blau, E., and Berardinelli, J. L.,** The role of peritoneal dialysis in Reye's syndrome: clinical diagnosis and treatment with peritoneal dialysis, *Pediatrics,* 53, 336, 1974.
16. **Roe, C. R., Schonberger, L. B., Gelbach, S. H., Wies, L. A., and Sidbury, J. B., Jr.,** Enzymatic alterations in Reye's syndrome prognostic implications, *Pediatrics,* 55, 119, 1975.
17. **Brown, R. E. and Madge, G. E.,** Hepatic degeneration and dysfunction in Reye's syndrome, *Am. J. Dig. Dis.,* 16, 1116, 1971.
18. **Boutros, A., Hoyt, J., Menezenes, A., and Bell, W.,** Management of Reye's syndrome. A rational approach to a complex problem, *Crit. Care Med.,* 5, 234, 1977.
19. **DeVivo, D. C. and Keating, J. P.,** Reye's syndrome, *Adv. Pediatr.,* 22, 175, 1976.
20. **Huttenlocher, P. R.,** Reye's syndrome: relation of outcome to therapy, *J. Pediatr.,* 80, 845, 1972.
21. **Lovejoy, F. H., Smith, A. L., Bresnan, M. J., Wood, J. N., Victor, D. I., and Adams, P. C.,** Clinical staging in Reye syndrome, *Am. J. Dis. Child.,* 128, 36, 1974.
22. **Shaywitz, B. A., Leventhal, J. M., Kramer, M. S., and Venes, J. L.,** Prolonged continuous monitoring of intracranial pressure in severe Reye's syndrome, *Pediatrics,* 59, 595, 1975.
23. **Aoki, Y. and Lombrosso, C. T.,** Prognostic value of electroencephalography in Reye's syndrome, *Neurology,* 23, 333, 1973.
24. **Engle, W. D., Baublis, J. V., Duff, T. E., Rosenberg, N. M., Tucker, R., and Kindt, C. W.,** Reye syndrome in Michigan, in *Reye's Syndrome II,* Crocker, J. F. S., Ed., Grune & Stratton, New York, 1979, 195.
25. **Marshall, L. F., Shapiro, H. M., Rauscher, A., and Kaufman, N. M.,** Pentobarbital therapy for intracranial hypertension in metabolic coma — Reye's syndrome, *Crit. Care Med.,* 6, 1, 1978.
26. **Venes, J. L., Shaywitz, B. A., and Spencer, D. D.,** Management of severe cerebral edema in the metabolic encephalopathy of Reye-Johnson syndrome, *J. Neurosurg.,* 48, 903, 1978.
27. **Trauner, D. A.,** Reye's syndrome, Elsevier, Amsterdam, 131, 133, 1982.
28. **Bradford, W. D. and Parker, J. C., Jr.,** Reye's syndrome. Possible causes and pathogenetic pathways, *Clin. Pediatr.,* 10, 148, 1971.
29. **Norman, M. C.,** Encephalopathy and fatty degeneration of the viscera in childhood. I. Review of cases at the hospital for sick children, Toronto (1954—1966), *Can. Med. Assoc. J.,* 99, 522, 1968.
30. **Baliga, R., Fleischman, L. E., Chang, Ch., Sarniak, A. D., Bidani, A., and Arcinue, E. L.,** Acute renal failure in Reye's syndrome, *Am. J. Dis. Child.,* 133, 1009, 1979.
31. **Gilboa, N.,** Reye syndrome and renal failure (letter to the Editor), *J. Pediatr.,* 95, 664, 1979.
32. **Mor, J., Susi, M., Kahan, E., Dawn, F., Teichberg, S., and McVicar, M.,** Acute renal failure in Reye's syndrome, *J. Pediatr.,* 94, 69, 1979.
33. **Schwartz, A. D.,** The coagulation defect in Reye's syndrome, *J. Pediatr.,* 78, 326, 1971.
34. **Tonsgard, J. H., Huttenlocher, P. R., and Thisted, R. A.,** Lactic acidemia in Reye's syndrome, *Pediatrics,* 69, 64, 1982.
35. **Hilty, M. D. and Romshe, C. A.,** Reye's syndrome and hyperaminoacidemia, *J. Pediatr.,* 84, 362, 1974.
36. **Romshe, C. A., Hilty, M. D., McClung, H. J., Kerzner, B., and Reiner, C. B.,** Amino acid pattern in Reye syndrome: comparison with clinically similar entities, *J. Pediatr.,* 98, 788, 1981.
37. **Huttenlocher, P. R. and Trauner, D. A.,** Reye's syndrome in infancy, *Pediatrics,* 62, 84, 1978.
38. **Jabbour, J. T., Howard, P. H., and Jacques, E. W.,** Encephalopathy and fatty degeneration of the liver and kidney, *JAMA,* 194, 1245, 1965.
39. **Curry, A. S., Guttman, H. A. N., and Price, D. F.,** An urinary pteridine in a case of liver failure, *Lancet,* 1, 885, 1966.
40. **Dvoráčková, I., Vortel, V., and Hroch, M.,** Encephalitic syndrome with fatty degeneration of viscera, *Arch. Pathol.,* 81, 240, 1966.
41. **Corey, L., Rubin, R. J., and Hattwick M. A. W.,** Reye's syndrome: clinical progression and evaluation of therapy, *Pediatrics,* 60, 708, 1977.
42. **Trauner, D. A.,** Treatment of Reye syndrome, *Ann. Neurol.,* 7, 2, 1980.
43. **Berman, W., Pizzi, F., Schut, L., Raphaely, R., and Holtzapple, P.,** The effects of exchange transfusion on intracranial pressure in patients with Reye syndrome, *J. Pediatr.,* 87, 887, 1975.
44. **Bobo, R. C., Schubert, W. K., Partin, J. C., and Partin, J. S.,** Reye's syndrome: treatment by exchange transfusion with special reference to the 1974 epidemic in Cincinnati, Ohio, *J. Pediatr.,* 87, 881, 1975.
45. **Glasgow, A. M. and Chase, H. P.,** Exchange transfusion to remove ammonia, *Am. J. Dis. Child.,* 129, 159, 1975.
46. **Lansky, L. L., Kalavsky, S. M., Brackett, C. E., Wallas, C. H., and Reis, R. L.,** Hypothermic total body washout and intracranial pressure monitoring in stage IV Reye syndrome, *J. Pediatr.,* 90, 639, 1977.
47. **Pegellow, C., Goldberg, R., Turkel, S., and Powers, D.,** Severe coagulation abnormalities in Reye syndrome, *J. Pediatr.,* 91, 413, 1977.
48. **Bourgeois, C. H., Olson, L., Comar, D., Evans, H., Keschamras, N., Cotton, R., Grossman, R., and Smith, T.,** Encephalopathy and fatty degeneration of the viscera: aclinico-pathologic analysis of 40 cases, *Am. J. Clin. Pathol.,* 56, 558, 1971.

49. **Castleman, B., McNeely, B. V., and Gellis, S. S.,** Case records of the Massachussetts general hospital, *N. Engl. J. Med.*, 276, 47, 1967.
50. **Fronstin, M. H., Moore, L. W., Ruffolo, E. H., and Hopper, G. S.,** Encephalopathy and fatty degeneration of the viscera: report of three cases in the same community, *Am. J. Clin. Pathol.*, 49, 704, 1968.
51. **Rahal, J. J. and Henle, G.,** Infectious mononucleosis and Reye's syndrome. A fatal case with studies for Epstein-Barr virus, *Pediatrics*, 46, 776, 1970.
52. **Bove, K. E., McAdams, A. J., Partin, J. C., Partin, J. S., Hug, G., and Schubert, W. K.,** The hepatic lesion in Reye's syndrome, *Gastroenterology*, 69, 685, 1975.
53. **Gilboa, N.,** Reye's syndrome and pancreatitis, *Am. J. Dis. Child.*, 134, 903, 1980.
54. **Pedal, I., Oehmicken, M., and Raff, G.,** Reye-syndrome mit pankreatitis und hypoxischer Hirnschädigung, *Dtsch. Med. Wochenschr.*, 109, 101, 1984.
55. **Partin, J. C., Schubert, W. K., and Partin, J. S.,** Mitochondrial ultrastructure in Reye's syndrome, *N. Engl. J. Med.*, 285, 1339, 1971.
56. **Partin, J. C., Partin, J. S., Schubert, W. K., and McLaurin, R. L.,** Brain ultrastructure in Reye's syndrome, *J. Neuropathol. Exp. Neurol.*, 34, 425, 1975.
57. **Partin, J. C., Bove, K., Partin, J. S., and Schubert, W. K.,** Liver and muscle ultrastructure in Reye's syndrome, in *Reye's Syndrome II*, Crocker, J. F. S., Ed., Grune & Stratton, New York, 1979, 217.
58. **Glasgow, A. M.,** Overview of metabolic abnormalities in Reye's syndrome, in *Reye's Syndrome IV*, Pollack, J. D., Ed., National Reye's Syndrome Foundation, Bryan, OH, 1985, 142.
59. **Thaler, M. M.,** Clinical and enzymatic indices of hepatic dysfunction in Reye's syndrome, in *Reye's Syndrome II*, Crocker, J. F. S., Ed., Grune & Stratton, New York, 1979, 115.
60. **Huttenlocher, P. R., Schwartz, A. D., and Klatskin, G.,** Reye's syndrome. Ammonia intoxication as a possible factor in the encephalopathy, *Pediatrics*, 43, 443, 1969.
61. **Fitzgerald, J. F., Clark, J. H., Angelides, A. G., and Wyllie, R.,** The prognostic significance of peak ammonia level in Reye syndrome, *Pediatrics*, 70, 997, 1982.
62. **Trauner, D. A., Sweetman, L., Holm, J., Kulovich, S., and Nyhan, W. L.,** Biochemical correlates of illness and recovery in Reye's syndrome, *Ann. Neurol.*, 2, 238, 1977.
63. **Glasgow, A. M., Cotton, R. B., and Dhiensire, K.,** Reye's syndrome. I. Blood ammonia and consideration of the nonhistologic diagnosis, *Am. J. Dis. Child.*, 124, 827, 1972.
64. **Thaler, M. M., Hoogenraad, N. J., and Boswell, M.,** Reye's syndrome due to a novel protein-tolerant variant of ornithin-trans-carbamylase deficiency, *Lancet*, 2, 438, 1974.
65. **Brown, T., Brown, H., Lansky, L., and Hig, G.,** Carbamyl phosphate sythetase and ornithin transcarbamylase in liver of Reye's syndrome patients, *N. Engl. J. Med.*, 291, 797, 1974.
66. **Rittenhouse, J., Mason, M., and Baublic, J. V.,** Amino acid ratios in Reye's syndrome, *Lancet*, 1, 105, 1973.
67. **Kang, E. S., Snodgrass, P. J., and Gerald, P. S.,** Ornithin transcarbamylase deficiency in the newborn infant, *J. Pediatr.*, 82, 642, 1973.
68. **Mamunes, P., DeVries, G. H., Miller, C. D., and David, R. B.,** Fatty acids in Reye's syndrome, *Pediatr. Res.*, 8, 436, 1974.
69. **Ogburn, P. L., Sharp, H., Lloyd-Still, J. D., Johnson, S. B., and Holman, R. T.,** Abnormal polyunsaturated fatty acids pattern of serum lipids in Reye's syndrome, *Proc. Natl. Acad. Sci. U.S.A.*, 79, 908, 1982.
70. **Trauner, D. A.,** Pathologic changes in a rabbit model of Reye's syndrome, *Pediatr. Res.*, 16, 950, 1982.
71. **Trauner, D. A. and Adams, H.,** Intracranial pressure elevations during octanoate infusion in rabbits: an experimental model of Reye's syndrome, *Pediatr. Res.*, 15, 1097, 1981.
72. **Roe, Ch. R., Millington, D. S., Maltby, D. A., and Bohan, T. P.,** Relative carnitine insufficiency in Reye's syndrome and related metabolic disorders, in *Reye's Syndrome IV.*, Pollack, J. D., Ed., National Reye's Syndrome Foundation, Bryan, OH, 1985, 201.
73. **Brown, R. E.,** Excessive free fatty acidemia and the encephalopathy in Reye's syndrome, *N. Engl. J. Med.*, 292, 1297, 1975.
74. **Robinson, B. H., Gall, D. G., and Cutz, E.,** Deficient activity of hepatic pyruvate dehydrogenase and pyruvate carboxylase in Reye's syndrome, *Pediatr. Res.*, 11, 279, 1977.
75. **Trauner, D. A.,** Reye's syndrome, *Curr. Probl. Pediatr.*, 12, 1, 1982.
76. **Faraj, B. A., Newman, S. L., Caplan, D. B., Ali, F. M., Camp, V. M., and Ahmann, P. A.,** Evidence for hypertyraminemia in Reye's syndrome, *Pediatrics*, 64, 76, 1979.
77. **Lloyd, K. G., Davidson, L., Price, K., McClung, H. J., and Gall, D. G.,** Catecholamine and octopamine concentrations in brains of patients with Reye syndrome, *Neurology*, 27, 985, 1977.
78. **Mitchell, R. A., Ram, M. L., Arcinue, E. L., and Chung, H. C.,** Comparison of cytosolic and mitochondrial hepatic enzyme alterations in Reye's syndrome, *Pediatr. Res.*, 14, 1216, 1980.
79. **Faraj, B. A., Caplan, D. B., Malveaux, E. J., Camp, V. M., and Farouk, M. A.,** Similarity between tyramin-induced neurotoxicity and the coma of Reye's syndrome, *J. Pharmacol. Exp. Ther.*, 226, 608, 1983.

80. **Green, H. L., Wilson, F. A., Glick, A. D., Dunn, G. D., and Kilroy, A. W.,** Hepatic ATP concentrations and glycolytic enzyme activities in Reye syndrome, *J. Pediatr.*, 89, 777, 1976.
81. **Robinson, B. H., Taylor, J., Cutz, E., and Gall, G. D.,** Reye's syndrome: preservation of mitochondrial enzymes in brain and muscle compared with liver, *Pediatr. Res.*, 12, 1045, 1978.
82. **Corey, L., Rubin, R. J., Hattwick, M. A. W., Noble, G. R., and Cassidy, E.,** A nationwide outbreak of Reye's syndrome. Its epidemiologic relationship to influenza B, *Am. J. Med.*, 61, 615, 1976.
83. **Morens, D. M., Sullivan-Bolyai, J. Z., Slater, J. E., Schonberger, L. B., and Nelson, D. B.,** Surveillance of Reye syndrome in the United States 1977, *Am. J. Epidemiol.*, 114, 406, 1981.
84. **Johnson, G. M., Scurletis, T. D., and Caroll, N. B.,** A study of 16 fatal cases of encephalitis-like disease in North Carolina children, *N.C. Med. J.*, 24, 464, 1963.
85. **Nelson, D. B., Hurwitz, E. S., Sullivan-Bolyai, J. Z., Morens, D. M., and Schonberger, L. B.,** Reye's syndrome in the United States in 1977—1978. A non-influenza B virus year, *J. Infect. Dis.*, 140, 436, 1979.
86. **Barrett, M. J., Hurwitz, E. S., Rogers, M. F., and Schonberger, L. B.,** The epidemiology of Reye's syndrome in the United States: the national Reye's syndrome surveillance system 1974—1984, in *Reye's Syndrome IV.*, Pollack, J. D., Ed., National Reye's Syndrome Foundation, Bryan, OH, 1985, 69.
87. **Hall, S. M and Bellman, M. H.,** Reye's syndrome in the British Isles. The British Paediatric Association — PHLS communicable disease surveillance centre joint surveillance scheme, in *Reye's Syndrome IV.*, Pollack, J. D., Ed., National Reye's Syndrome Foundation, Bryan, OH, 1985, 32.
88. **Yamashita, F., Ono, E., Kimura, A., and Yoshida, I.,** Reye's syndrome in Asia, in *Reye's Syndrome IV.*, Pollack, J. D., Ed., National Reye's Syndrome Foundation, Bryan, OH, 1985, 47.
89. **Abruzzi, W. A.,** Varicella encephalitis, *N.Y. J. Med.*, 61, 3912, 1961.
90. **Barr, R., Glass, I. H. T., and Chawla, G. S.,** Reye's syndrome: massive fatty metamorphosis of the liver with acute encephalopathy, *Can. Med. Assoc. J.*, 98, 1038, 1968.
91. **Blair, A. W., Jamieson, W. M., and Smith, G. R.,** Complications and death in chickenpox, *Br. Med. J.*, 2, 981, 1965.
92. **Brown, R. E., Madge, G. E., and Schiller, H. M.,** Observations on the pathogenesis of Reye's syndrome, *South. Med. J.*, 64, 942, 1971.
93. **Glasgow, A. M. and Gold, M. B.,** Interval between varicella and Reye's syndrome, *Am. J. Dis. Child.*, 133, 653, 1979.
94. **Holmes, G. L.,** Reye's syndrome. A review of six cases, *Va. Med. Mon.*, 30, 466, 1977.
95. **Jenkins, R., Dvorak, A., and Patrick, J.,** Encephalopathy and fatty degeneration of the viscera associated with chickenpox, *Pediatrics*, 39, 769, 1967.
96. **Linnemann, C. C., Jr., Shea, L., Partin, J. C., Schubert, W. K., and Schiff, G. M.,** Reye's syndrome: epidemiology and viral studies 1963—1974, *Am. J. Epidemiol.*, 101, 517, 1975.
97. **Nader, P. and Leonards, R.,** Varicella and hypoglycemia with recovery, *Am. J. Dis. Child.*, 110, 678, 1965.
98. **Hurwitz, E. S. and Goodman, R. A.,** A cluster of cases of Reye syndrome associated with chickenpox, *Pediatrics*, 70, 901, 1982.
99. **Stechenberg, B. W., Keating, J. P., Kosloc, S., Schechter, M., Chang, M., Haymond, M. W., and Feigin, R. D.,** Epidemiologic investigation of Reye syndrome, *J. Pediatr.*, 87, 234, 1975.
100. **Corey, L. and Rubin, J.,** Reye's syndrome 1974: an epidemiological assessment, in *Reye's Syndrome*, Pollack, J. D., Ed., Grune & Stratton, New York, 1975, 179.
101. **Golden, G. S. and Duffel, D.,** Encephalopathy and fatty change in the liver and kidney, *Pediatrics*, 36, 67, 1965.
102. **Hochberg, F. H., Nelson, K., and Jansen, W.,** Influenza type B — related encephalopathy, *JAMA*, 231, 1975.
103. **LaMontagne, J. R.,** Summary of a workshop in influenza B viruses and Reye syndrome, *J. Infect. Dis.*, 142, 452, 1980.
104. **Reynolds, D. W., Riley, H. D., LaFont, D. S., Vorse, H., Stout, L. C., and Carpenter, R. L.,** An outbreak of Reye's syndrome associated with influenza B, *J. Pediatr.*, 80, 429, 1972.
105. **Linnemann, C. C., Kaufman, C. A., Shea, L., Schiff, G., Partin, J. C., and Schubert, W. K.,** Association of Reye's syndrome with viral infection, *Lancet*, 1, 179, 1974.
106. **Morgan, D. and Noble, G. R.,** Reye syndrome and influenza, *Lancet*, 1, 807, 1977.
107. **Norman, M. G., Lowden, J. A., Hill, D. E., and Bannatyre, R. M.,** Encephalopathy and fatty degeneration of the viscera in childhood. II. Report of a case with isolation of influenza B virus, *Can. Med. Assoc. J.*, 99, 549, 1968.
108. **Ruben, F. L. and Michaels, R. H.,** Reye syndrome with associated influenza A and B infection, *JAMA*, 234, 410, 1975.
109. **Ruben, F. L., Streiff, E. J., Meal, M., and Michaels, R. H.,** Epidemiologic studies of Reye's syndrome: cases seen in Pittsburgh, October 1973—April 1975, *Am. J. Publ. Health*, 66, 1096, 1976.
110. **Wilson, R., Miller, J., Greene, H., Rankins, R., Lumeng, L., Gordon, D., Nelson, D., and Noble, G.,** Reye's syndrome in three siblings: association with type A influenza infection, *Am. J. Dis. Child.*, 134, 1032, 1980.
111. **Monto, A. S., Ceglarek, J. P., and Hayner, N. S.,** Liver function abnormalities in the course of a type A (H_1N_1) influenza outbreak. Relation to Reye's syndrome, *Am. J. Epidemiol.*, 114, 750, 1981.

112. **Davis, L. E. and Kornfeld, M.,** Influenza A virus and Reye's syndrome in adults, *J. Neurol. Neurosurg. Psychiatry,* 43, 516, 1980.
113. **Hall, B. D., Hughes, W. T., and Kmetz, D.,** Reye's syndrome: an association with influenza A infection, *J. Ky. Med. Assoc.,* 67, 269, 1969.
114. **Halsey, N. A., Hurwitz, E. S., Meiklejohn, G., Todd, W. A., Edell, T., Todd, J. K., and McIntosh, K.,** An epidemic of Reye syndrome associated wiht influenza A (H_1N_1) in Colorado, *J. Pediatr.,* 97, 535, 1980.
115. **Nicholls, S., Gill, D., and Craske, J.,** Reye's syndrome associated with acute tubular necrosis, *Arch. Dis. Child.,* 50, 960, 1975.
116. **Noble, G. R., Corey, L., and Rubin, R. J.,** Virologic components of Reye's syndrome, in *Reye's Syndrome I.,* Pollack, J. D., Ed., Grune & Stratton, New York, 1975, 189.
117. **Partin, J. C., Schubert, W. K., Partin, J. S., Jacobs, R., and Saalfeld, K.,** Isolation of influenza virus from liver and muscle biopsy specimens from a surviving case of Reye's syndrome, *Lancet,* 1, 599, 1976.
118. **Bachman, H. J. and Haupt, M.,** Reye syndrome: encephalopathie mit fettiger Degeneration der viszeralen Organen, *Klin. Pediatr.,* 185, 313, 1973.
119. **Brown, R. E., Schiller, H. M., Madge, G. E., and Still, W. J. S.,** Reye's syndrome: ultrastructural features and observations on pathogenesis, *South. Med. J.,* 63, 1230, 1970.
120. **Kaul, A., Cohen, M. E., Broffman, G., Fischer, J., Jenis, E. H., and Ogra, P. L.,** Reye-like syndrome associated with coxsackie B_2 virus infection, *J. Pediatr.,* 94, 67, 1979.
121. **Utian, H. L., Wagner, J. M., and Sichel, R. J. S.,** White liver disease, *Lancet,* 2, 1043, 1964.
122. **Edwards, K. M., Bennett, S. R., Garner, W. L., Bratton, D. D., Glick, A. D., Greene, H. L., and Wright, P. F.,** Reye's syndrome associated with adenovirus infections in infants, *Am. J. Dis. Child.,* 139, 343, 1985.
123. **Dvořáčková, I. and Hroch, H.,** Einige epidemiologische und virologische Bemerkungen zum enzephalitischen Syndrom mit Organenverfettung, *Zentralbl. Bakteriol. Parasit. Infekt. Hyg.,* 206, 421, 1968.
124. **Ladisch, S., Lovejoy, F. H., Hierholzer, J. C., Oxman, M. N., Strieder, D., Vawter, G. F., Finer, N., and Moore, M.,** Extrapulmonary manifestations of adenovirus toxin, *J. Pediatr.,* 95, 348, 1979.
125. **Daugherty, C. C., Heubi, J. E., Edwards, K. M., and Greene, H.,** Reye's syndrome associated with adenovirus infections, *Am. J. Dis. Child.,* 139, 1076, 1985.
126. **Brown, J. H.,** Reye's syndrome associated with adenovirus type 3 infection, *Med. J. Aust.,* 2, 873, 1974.
127. **Cullity, G. and Kakulas, B. A.,** Encephalopathy and fatty degeneration of the viscera: an evaluation, *Brain,* 93, 77, 1970.
128. **Chalhub, E. G., DeVivo, D. C., Keating, J. P., Haymond, M. W., and Feigin, R. D.,** Reye syndrome complicated by a generalized herpes simplex type I infection, *J. Pediatr.,* 98, 73, 1981.
129. **Becroft, D. M. O.,** Syndrome of encephalopathy and fatty degeneration of the viscera in New Zealand children, *Br. Med. J.,* 2, 135, 1966.
130. **Dorman, J. M., Glick, T. H., Shannon, D. C., Galdabini, J., and Walker, W. A.,** Complications of infectious mononucleosis. A fatal case in a 2 year-old child, *Am. J. Dis. Child.,* 128, 238, 1974.
131. **Fleischer, G., Schwartz, J., and Lennette, E.,** Primary Epstein-Barr virus infection in association with Reye syndrome, *J. Pediatr.,* 97, 935, 1980.
132. **Powell, H. C., Rosenberg, R. N., and McKellan, B.,** Reye's syndrome: isolation of *Parainfluenza* virus, *Arch. Neurol.,* 29, 135, 1973.
133. **Joske, R. A., Keall, D. D., Leak, P. J., Stanley, N. F., and Walters, M. N.,** Hepatitis encephalitis in humans with reovirus infection, *Arch. Intern. Med.,* 113, 811, 1964.
134. **Griffin, N., Keeling, J. W., and Tomlinson, H. H.,** Reye's syndrome associated with respiratory syncytial virus infection, *Arch. Dis. Child.,* 54, 74, 1979.
135. **Bell, W. E.,** Reye's syndrome: an association with type 1 virion like poliovirus, *Arch. Neurol.,* 30, 304, 1974.
136. **Sherman, P. E., Michaels, R. H., and Kenny, F. M.,** Acute encephalopathy complicating rubeolla, *JAMA,* 192, 675, 1965.
137. **Morens, D. M., Halsey, N. A., Schonberger, L. S. and Baublis, J. V.,** Reye syndrome associated with vaccination of live virus vaccines, *Clin. Pediatr.,* 18, 42, 1979.
138. **Tang, T. T., Siegesmund, K. A., Sednak, G. V., Casper, J. T., Varma, R., and McCreadie, S. R.,** Reye's syndrome: a correlated electronmicroscopic, viral and biochemical observation, *JAMA,* 232, 1339, 1975.
139. **Morens, D. and Noble, G. R.,** Reye's syndrome and influenza, *Lancet,* 2, 807, 1977.
140. **Turel, A. P., Levinsohn, M. W., Derakhshan, I., and Gutierrez, Y.,** Reye's syndrome and cerebellar intracytoplasmic inclusion bodies, *Arch. Neurol.,* 32, 624, 1975.
141. **Iancu, T. C., Mason, W. H., and Neustein, H. B.,** Ultrastructural abnormalities of liver cells in Reye's syndrome, *Hum. Pathol.,* 8, 421, 1977.
142. **Collins, D. N. and Gilbert, E. F.,** Glycogen complexes in muscle in Reye's syndrome simulating virus-like particles, *Lab. Invest.,* 36, 91, 1977.
143. **Lewinski, U. H. and Djaldetti, M.,** Ultrastructural alterations of the lymphocytes from a patient with Reye's syndrome, *J. Submicrosc. Cytol.,* 13, 697, 1981.
144. **Davis, L. E., Cole, L. L., Lockwood, S. J., and Kornfeld, M.,** Experimental influenza B virus toxicity in mice. A possible model for Reye's syndrome, *Lab. Invest.,* 48, 140, 1983.

145. **Younkin, B. and Gudzimovicz, B.,** The viral mechanism of Reye's syndrome, *Med. Hypotheses,* 14, 161, 1984.
146. **Hanson, P. A. and Urizar, R. E.,** Ultrastructural lesions of muscle and immunofluorescent deposits in vessels in Reye's syndrome. A preliminary report of serial muscle biopsies, *Ann. Neurol.,* 1, 431, 1977.
147. **Tang, T. T., Harb, J. M., Grossberg, S. E., Sedmak, G. V., and Murphy, J. V.,** Leucocyte tubuloreticular inclusions in Reye's syndrome, *Arch. Pathol. Lab. Med.,* 109, 543, 1985.
148. **Kang, E. S., Solomon, S. S., Duckworth, W. C., Burghen, G. A., and Schwenzer, K. S.,** Abnormal pancreatic endocrine function in Reye's syndrome survivors and their relatives: a preliminary report, *Pediatr. Res.,* 13, 870, 1979.
149. **Thaler, M. M., Bruhn, F. W., Applebaum, M. N., and Goodman, J.,** Reye's syndrome in twins, *J. Pediatr.,* 77, 638, 1970.
150. **Hilty, M. D., McClung, J. H., Haynes, R. E., Romshe, C. A., and Sherard, E. S.,** Reye syndrome in siblings, *J. Pediatr.,* 94, 576, 1979.
151. **Brown, T., Hug, G., Lansky, L., Bove, K., Scheve, A., Ryan, M., Brown, H., Schubert, W. K., Partin, J. C., and Lloyd-Still, J.,** Transiently reduced activity of carbamyl phosphate synthetase and ornithin transcarbamylase in liver of children with Reye's syndrome, *N. Engl. J. Med.,* 294, 861, 1976.
152. **Yokoi, T., Henke, K., Funabashi, T., Hayashi, R., Suzuki, Y., Taniguchi, N., Hosoya, N., and Saheki, T.,** Partial ornithin transcarbamylase deficiency simulating Reye's syndrome, *J. Pediatr.,* 99, 929, 1981.
153. **Chapoy, P., Angelini, C., and Cederbaum, S.,** Déficit systémique en carnitine. Place dans le sydrome de Reye, *Nouv. Presse Med.,* 10, 499, 1981.
154. **DelValle, J. A., Garcia, H. J., Merinero, B., Perez-Cerda, C., Roman F., Jimenez, A., and Ugarto, M.,** A new patient with dicarboxylic acidurie suggestive of medium-chain acyl-CoA dehydrogenase deficiency presenting as Reye's syndrome, *J. Inherited Metab. Dis.,* 7, 62, 1984.
155. **Stanley, C. A., Hale, D. E., Coates, P. M., Hall, C. L., Corkey, B. E., Yang, W., Kelley, R. I., Gonzales, E. L., Williamson, J. R., and Baker, L.,** Medium-chain acyl-CoA dehydrogenase deficiency in children with non-ketotic hypoglycemia and low carnitine levels, *Pediatr. Res.,* 17, 877, 1983.
156. **Aprille, J. R.,** Reye's syndrome: patient serum alters mitochondrial function and morphology *in vitro, Science,* 197, 908, 1977.
157. **Aprille, J. R.,** A serum factor in Reye's syndrome, *J. Natl. Reye's Syndr. Found.,* 1, 37, 1980.
158. **Aprille, J. R., Austin, J., Castello, C. E., and Royal, N.,** Identification of the Reye's syndrome "serum factor", *Biochem. Biophys. Res. Commun.,* 94, 381, 1980.
159. **Scott, H. H.,** On the vomiting sickness of Jamaica, *Am. Trop. Med. Parasit.,* 10, 1, 1917.
160. **Glasgow, A. M. and Chase, H. P.,** Production of the features of Reye's syndrome in rats with 4-pentenoic acid, *Pediatr. Res.,* 9, 133, 1975.
161. **Gerber, N., Dickinson, R. G., Harland, R. C., Lynn, R. K., Houghton, D., and Antonias, J. I.,** Reye-like syndrome associated with valproic acid therapy, *J. Pediatr.,* 95, 142, 1979.
162. **Sugimoto, T., Nishida, N., Aysuhara, A., Ono, A., Sakane, Y., and Matsumura, T.,** Reye-like syndrome associated with valproic acid, *Brain Dev.,* 5, 334, 1983.
163. **Shield, L. K., Coleman, T. L., and Markesbery, W. R.,** Methyl bromide intoxication: neurologic features, including simulation of Reye syndrome, *Neurology,* 27, 959, 1977.
164. **Fox, D. W., Hart, M. C., Bergeson, P. S., Jarrett, P. B., Stillman, A. E., and Huxtable, R. J.,** Pyrrolozidine (Senecio) intoxication mimicking Reye syndrome, *J. Pediatr.,* 93, 980, 1978.
165. **Sinniah, D. and Baskaran, G.,** Margosa oil poisoning as a cause of Reye's syndrome, *Lancet,* 1, 487, 1981.
166. **Jimenez, J. F., Brown, A. L., Arnold, W. C., and Byrne, W. J.,** Chronic camphor ingestion mimicking Reye's syndrome, *Gastroenterology,* 84, 394, 1983.
167. **Heick, H. M. C., Shipman, R. T., Norman, M. G., and James, W.,** Reye-like syndrome associated with use of insect repellent in a presumed heterozygote for ornithin carbamyl transferase deficiency, *J. Pediatr.,* 97, 471, 1980.
168. **Mortimer, E. A. and Lepow, M. L.,** Varicella with hypoglycemia possibly due to salicylates, *Am. J. Dis. Child.,* 103, 583, 1962.
169. **Giles, H. M.,** Encephalopathy and fatty degeneration of the viscera (letter), *Lancet,* 1, 1075, 1965.
170. **Bernstein, B. H., Singen, B. H., King, K. K., and Hanson, V.,** Aspirin induced hepatotoxicity and its effect on juvenile rheumatoid arthritis, *Am. J. Dis. Child.,* 131, 659, 1977.
171. **Atwood, S. J.,** The laboratory in the diagnosis and management of acetaminophen and salicylate intoxications, *Pediatr. Clin. North Am.,* 27, 871, 1980.
172. **Brown, R. D. and Wilson, J. T.,** Aspirin consumption and severity of Reye's syndrome, *Pediatrics,* 71, 293, 1983.
173. **Partin, J. S., Daugherty, C. C., McAdams, A. J., Partin, J. C., and Schubert, W. K.,** A comparison of ultrastructure in salicylate intoxication and Reye's syndrome, *Hepatology,* 4, 687, 1984.
174. **Yoshida, I., Yamashita, F., Okada, S., and Horihoshi, T.,** Simulated Reye's syndrome and salicylate therapy, *Acta Pediatr. Scand.,* 73, 562, 1984.
175. **Iancu, T and Ellran, E.,** Ultrastructural changes in aspirin hepatotoxicity, *Am. J. Clin. Pathol.,* 66, 570, 1976.

176. **Starko, K. M., Ray, C. G., Dominguez, L. B., Stromberg, W. L., and Woodall, D. F.,** Reye's syndrome and salicylate use, *Pediatrics,* 66, 859, 1980.
177. **Waldman, R. J., Hall, W. N., McGee, H., and van Amburg, G.,** Aspirin as a risk factor in Reye's syndrome, *JAMA,* 247, 3089, 1982.
178. **Halpin, T. J., Holtzhauer, F. J., Campbell, R. J., Hall, L. J., Correa–Villasener, A., Lanese, R., Rice, J., and Hurwitz, E. S.,** Reye's syndrome and medication use, *JAMA,* 248, 687, 1982.
179. **Clark, J. H. and Fitzgerald, J. F.,** Doubts relationship of salicylate and Reye's syndrome, *Pediatrics,* 68, 467, 1981.
180. **Gall, D. G., Barker, G., and Cutz, E.,** Doubts relationship of salicylate and Reye's syndrome (to the editor), *Pediatrics,* 68, 467, 1981.
181. **Daniels, S. R., Greenberg, R. S., and Ibrahim, M. A.,** Scientific uncertainties in the studies of salicylate use and Reye's syndrome, *JAMA,* 249, 1311, 1983.
182. **Partin, J. S., Partin, J. C., Schubert, W. K., and Hammond, J. G.,** Serum salicylate concentrations in Reye's disease. A study of 130 biopsy proven cases, *Lancet,* 1, 191, 1982.
183. **Bailey, J. M., Low, C. E., and Pupillo, M. B.,** Reye's syndrome and aspirin use: a possible immunological relationship, *Prostaglandins Leukotrienes Med.,* 8, 211, 1982.
184. **Rodgers, G. C., Jr.,** Analgesies and Reye's syndrome: fact or fiction?, in *Reye's Syndrome IV,* Pollack, J. D., Ed., National Reye's Syndrome Foundation, Bryan, OH, 1985, 117.
185. **Crocker, J. F. S., Rozee, K. R., Ozere, R. L., Digout, S. C., and Hutzinger, O.,** Insecticide and viral interaction as a cause of fatty visceral changes and encephalopathy in the mouse, *Lancet,* 2, 22, 1974.
186. **Colon, A. R., Pardo, V., and Sandberg, D. H.,** Experimental Reye's syndrome induced by viral potentiation of chemical toxin, in *Reye's Syndrome,* Pollack, J. D., Ed., Grune & Stratton, New York, 1975, 199.
187. **Corey, L., Rubin, R. J., Bregman, D., and Gregg, M. B.,** Diagnostic criteria for influenza B associated Reye's syndrome: clinical vs. pathologic criteria, *Pediatrics,* 60, 702, 1977.
188. **Crocker, J. F. S., Ozere, R. L., Safe, S. H., Digout, S. C., Rozee, K. R., and Hutzinger, O.,** Lethal interaction of ubiquitous insecticide carriers with virus, *Science,* 192, 1351, 1976.
189. **Colon, A. R., Ledesma, F., Pardo, V., and Sandberg, D. H.,** Viral potentiation of chemical toxins in the experimental syndrome of hypoglycemia, encephalopathy, and visceral fatty degeneration, *Am. J. Digest. Dis.,* 19, 1091, 1974.
190. **Hug, G., Bosken, J., Bove, K., Linnemann, C. C. J., and McAdams, L.,** Reye's syndrome simulacra in liver of mice after treatment with chemical agents and encephalomyocarditis virus, *Lab. Invest.,* 45, 89, 1981.
191. **Olson, L. C., Bourgeois, C. H., Cotton, R. B., Harikul, S., Grossman, R. A., and Smith, T. J.,** Encephalopathy and fatty degeneration of the viscera in Northeastern Thailand. Clinical syndrome and epidemiology, *Pediatrics,* 47, 707, 1971.
192. **Bourgeois, C. H.,** Encephalopathy and fatty viscera: a possible response to acute aflatoxin poisoning, in *Reye's Syndrome,* Pollack, J. D., Ed., Grune & Stratton, New York, 1975, 131.
193. **Bourgeois, C. H., Shank, R. C., Grossman, R. A., Johnson, D. O., Wooding, W. L., and Chandavimol, P.,** Acute aflatoxin B_1 toxicity in the macaque and its similarities to Reye's syndrome, *Lab. Invest.,* 24, 206, 1971.
194. **Shank, R. C., Bourgeois, C. H., Keschamras, N., and Chandavimol, P.,** Aflatoxins in autopsy specimens from Thai children with an acute disease of unknown etiology, *Food Cosmet. Toxicol.,* 9, 501, 1971.
195. **Becroft, D. M. O. and Webster, D. R.,** Aflatoxins and Reye's disease, *Br. Med. J.,* 4, 117, 1972.
196. **Dvořáčková, I., Žilková, J., Brodský, F., and Cerman, J.,** Aflatoxin and liver damage with encephalopathy, *Sb. Věd. Pr. Lek. Fak. Karlovy Univ. Hradci Kralove,* 15, 521, 1972.
197. **Dvořáčková, I., Brodsky, F., and Cerman, J.,** Aflatoxin and encephalitic syndrome with fatty degeneration of viscera, *Nutr. Rep. Int.,* 10, 89, 1974.
198. **Dvořáčková, I., Kusák, V., Veselý, D., Veselá, D., and Nesnídal, P.,** Aflatoxin and encephalopathy with fatty degeneration of viscera (Reye), *Ann. Nutr. Aliment.,* 31, 977, 1977.
199. **Chaves-Carballo, E., Ellefson, R. D., and Gomez, M. R.,** An aflatoxin in the liver of a patient with Reye-Johnson syndrome, *Mayo Clin. Proc.,* 51, 48, 1976.
200. **Hogan, G. R., Ryan, N. J., and Hayes, A. W.,** Aflatoxin B and Reye's syndrome, *Lancet,* 1, 561, 1978.
201. **Ryan, N. J., Hogan, G. R., Hayes, A. W., Ungar, P. D., and Siraj, M. Y.,** Aflatoxin B_1: its role in the etiology of Reye's syndrome, *Pediatrics,* 64, 71, 1979.
202. **Nelson, D. B., Kimbrough, R. D., Landrigan, P. J., Hayes, A. W., Yang, G. C., and Benanides, J.,** Aflatoxin and Reye's syndrome. A case control study, *Pediatrics,* 66, 865, 1980.
203. **Rogan, W. J., Yang, G. C., and Kimbrough, R. D.,** Aflatoxin and Reye's syndrome: a study of liver from deceased cases, *Arch. Environ. Health,* 40, 91, 1985.
204. **Hayes, A. W., Unger, P. D., and Stoloff, L.,** Occurrence of aflatoxin in hypoallergenic milk substitutes, *J. Food Prot.,* 41, 494, 1978.
205. **Wray, B. B.,** Aflatoxin and Reye's syndrome: a case control study, *Pediatrics,* 68, 473, 1981.
206. **Friend, M. and Trainer, D. O.,** Experimental DDT-duck hepatitis virus interaction studies, *J. Wildl. Manage.,* 38, 887, 1974.

207. **Morris, O. N.,** Long-term effects of aeriel applications of virus–fenithrothion combinations against the spruce budworm, *Chorisoneura fumiferana* (Lepidoptera: Torticidae), *Can. Entomol.,* 109, 9, 1977.
208. **Gabliks, J. and Friedman, L.,** Effects of insecticides on mammalian cells and virus infections, *Proc. N.Y. Acad. Sci.,* 160, 254, 1969.
209. **Rozee, K. R., Lee, S. H. S., Crocker, J. F. S., and Safe, S.,** Enhanced virus replication in mammalian cells exposed to commercial emulsifiers, *Appl. Environ. Microbiol.,* 35, 297, 1978.
210. **Pollack, J. D.,** Models of chemical and virus interaction and their relation to a multiple etiology of Reye's syndrome, in *Reye's Syndrome II,* Crocker, J. F. S., Ed., Grune & Stratton, New York, 1979, 341.
211. **Mullen, P. W.,** Immunopharmacological considerations in Reye's syndrome: a possible xenobiotic initiated disorder?, *Biochem. Pharmacol.,* 27, 145, 1978.
212. **Rozee, K. R., Lee, S. H. S., Crocker, J. F. S., Digout, S., and Arcinue, E.,** Is a compromised interferon response an etiologic factor in Reye's syndrome?, *Can. Med. Assoc. J.,* 126, 798, 1982.

Chapter 3

A FOLLOW-UP STUDY OF REYE'S SYNDROME IN CZECHOSLOVAKIA

I. INTRODUCTION

The first cases of Reye's syndrome (RS) were observed at the department of Pathology in Hradec Králové since the year 1958 and first reported at the Congress of Czechoslovak pathologists on September 27, 1963, under the title of "Encephalitic Syndrome with fatty Degeneration of the Liver."[1]

During the first 13 years (1958 to 1970), the study concerned patients with RS who had died and been autopsied at the hospital in Hradec Králové. Their clinical, morphological, and epidemiological data are reviewed in the first part of this chapter. Later, the similarity between RS and the disease described by Bourgeois et al.[2] and Olson et al.[3] in Thai children in whom aflatoxin had been recovered from the tissues stimulated an interest in aflatoxin and its possible role in the etiology of this syndrome. A systematic 15-year study has been carried out to clarify the possible etiological relationship between aflatoxin and this children's disease. The results of this study are the subject of the second part of this chapter.

II. REYE'S SYNDROME CASES IN THE YEARS 1958 TO 1970

The series consisted of 19 children, 13 boys and 6 girls. The youngest was 2 months and the eldest $10^1/_2$ years old. Twelve of them were children under 1 year of age. The disease occurred sporadically throughout the year, somewhat more often in the spring and summer.

A. CLINICAL AND LABORATORY DATA

Frequency of the main clinical and laboratory findings is shown in Tables 7 and 8. In all but three children, there was an antecedent mild infection of the upper respiratory tract with fever, cough, and/or rhinorrhea. Two infants had gastrointestinal symptoms and two had been vaccinated 7 d prior to the onset of prodromal symptoms; one had received live oral polio virus, and the other had been inoculated with a smallpox vaccine. Seven patients had a rash of short duration. The children did not seem seriously ill, and there was a short period of apparent recovery before the more serious phase of the illness supervened. After several hours up to 3 d there was a fairly abrupt clinical deterioration associated with severe vomiting, followed by depressed sensorium, generalized convulsions, increased or variable muscle tone and tendon reflexes, irregular respiration, hyperpnea, terminal hyperpyrexia, and a coma. Hepatomegaly was found in 15 out of the 19 patients. Death usually occurred within 3 to 10 d. The treatment included antibiotics, steroids, and anticonvulsants. Three children received a single dose of 1 to 2 tablets of Acylpyrin (1 tablet contains 100 mg of acetylosalicylic acid). In all cases the clinical diagnosis was acute encephalitis; only in two cases hepatoencephalitic syndrome was diagnosed.

Laboratory investigations could not be performed in all cases, since most of the children died shortly after admission to the hospital. The available laboratory findings are presented in Table 8.

Neutrophilic leukocytosis was commonly present. Serum transaminases were elevated in the five patients in whom the measurements were made. The levels of AST ranged from 3.50 to 6.93 mkat/l and those of ALT ranged from 3.31 to 6.31 mkat/l. The blood sugar level was extremely low in five out of nine patients in whom it was measured (1.0 to 2.9 µmol/l). The blood urea was elevated in five out of seven investigated patients. The total of proteins and γ-globulins was

TABLE 7
Clinical Symptoms of Reye's Cases 1958—1970

No.	Sex	Age[a]	Month of onset	Antecedent illness URI	Antecedent illness GIT	Antecedent illness Vaccine	Fever	Rash	Vomiting	Respiratory alterations	Convulsions	Hepatomegaly	Death in days	Clinical diagnosis[b]	Therapy[c]
1	M	10 yr	IV	+			+		+	+	+	+	7	Enc.	Antib.
2	F	5 yr	VI		+			+	+	+	+	+	4	Enc.	Antib.
3	F	3 yr	VII	+			+	+	+	+	+	+	3	Tumor	Antib.
4	F	6 m	XI	+			+		+	+	+	+	4	Enc.	Antib.
5	M	5 m	I	+			+		+	+	+	+	6	Enc.	Acylpyr.
6	M	6 m	VI	+		Polio	+		+	+	+	+	6	Enc.	Antib.
7	M	3 m	VII	+			+		+	+	+	+	10	Enc.	Antib.
8	M	2 m	III	+			+	+	+	+	+	+	4	Enc.	Acylpyr.
9	M	4 m	VIII	+			+		+	+	+		5	Enc.	Acylpyr.
10	F	2.5 m	VIII	+			+		+	+	+	+	6	Enc.	Antib.
11	M	3 m	III		+		+		+	+	+	+	6	Enc.	Antib.
12	M	4 yr	III	+			+		+	+	+		6	Enc.	Antib.
13	F	4 yr	IX	+			+		+	+	+		6	Enc.	Antib.
14	M	7 m	IX			Smallpox	+	+	+	+	+		4	Enc.	Antib.
15	M	7 m	XII	+			+	+	+	+	+		6	Enc.	Antib.
16	F	3 yr	VIII	+			+	+	+	+	+	+	7	Enc.	Antib.
17	M	1 yr, 7 m	IX		+		+		+	+	+	+	3	Enc.	Antib.
18	M	6 m	I	+			+		+	+	+		3	Enc.	Antib.
19	M	3 m	VI	+			+		+	+	+	+	4	Enc.	Antib.
												+	7	Enc.	Antib.

Note: URI = upper respiratory infection. GIT = gastrointestinal symptoms.

[a] yr = years, m = months.
[b] Enc. = encephalitis.
[c] Antib. = antibiotics.

TABLE 8
Laboratory Data in Reye's Cases 1958—1970

No.	Blood								CSF			Urine
	WBC (10⁹/l)	TP (g/l)	γ-Globulin	Urea (mmol/l)	Prothromb. (time/s)	AST (μkat/l)	ALT (μkat/l)	Blood sugar (mmol/l)	Sugar (mmol/l)	Protein		
1	13	71	0.174	2.3	16.3			5.6	3.8		
2	11							4.2			
3	15										
4	31	47	0.173	44	15.4			4.3	2.9		
5	13										
6	22					4.70	5.63			+	
7	27	56	0.170	75	16.2	3.50	3.31	2.1	0.3		
8	10	42	0.167	64	14.8	6.93	6.31	1.1	0.8		
9	8	46	0.090	58	15.7			1.0	0.2		
10	7	61	0.093	73	17.3						
11	20										
12	19					5.31	6.21	3.7	1.3		
13	17										Hydroxypteridines
14				28	16.2						
15	15					5.11	5.73	2.9	0.5		
16	16										
17	20										
18					14.3			4.3	2.3	+	
19	10										

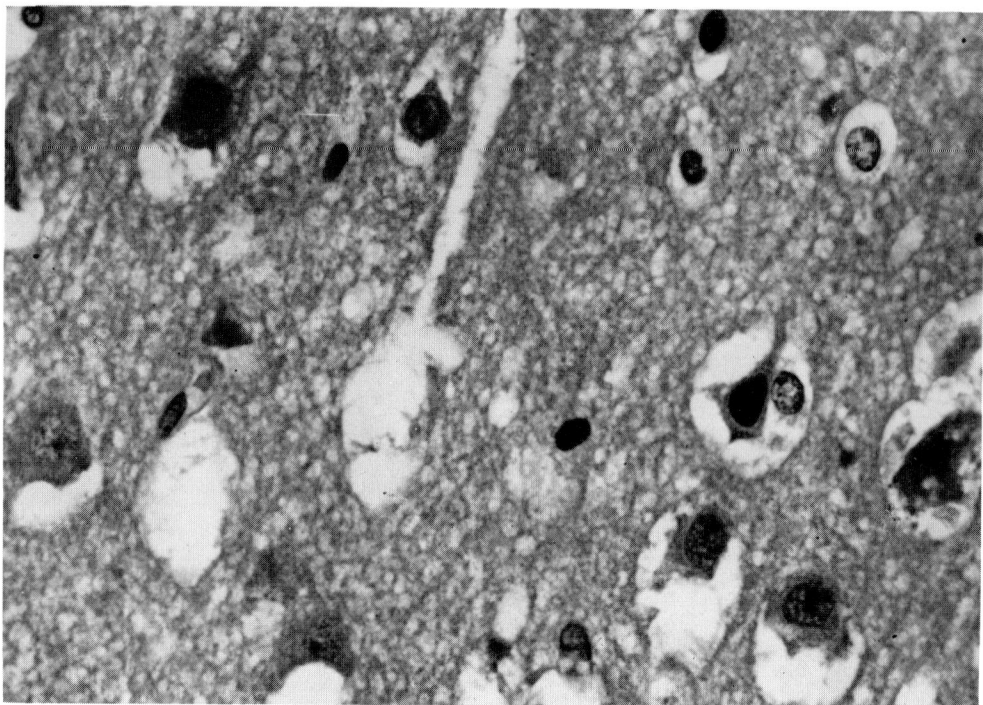

FIGURE 2. Pericellular edema and necrosis of isolated ganglion cells in the cerebral cortex (hematoxylin and eosin, ×400).

mildly depressed in three out of six children. The prothrombin time was prolonged (13.3 to 14.8 per second) in all eight patients in whom it was measured.

Examination of the cerebrospinal fluid revealed slight protein levels in two out of ten investigated cases. Extreme low sugar concentration (0.2 to 1.8 μmol/l) were found in five patients with concomitant hypoglycemia.

Tests performed on the urine for the presence of sugar, protein, and acetone were negative. In one patient the urine contained 7-hydroxypteridines.

B. MORPHOLOGICAL FEATURES
1. Gross Pathological Findings

The gross pathological findings were cerebral edema, enlarged fatty liver, pale kidneys and myocardium, hyperplasia of the lymph nodes, particularly of the mesentery, and hypoplasia of the adrenal glands in some instances. In the majority of cases, the stomach contained a small quantity of dark fluid colored by blood.

2. Histological Findings

The pathological changes found in the brain was spongioform edema of the cortex, particularly of the frontal lobes and the molecular and subcortical layers of the cerebellum. The neurons throughout the cortex were swollen with vacuolated plasma, and some contained sudanophilic granules in frozen sections; other neurons were shrunken, deeply staining, and some were necrotic (Figure 2). In all regions of the brain, the astrocytes were greatly swollen with pale nuclei corresponding to Alzheimer's glia type II (Figure 3). The Robin-Virchow's spaces contained numerous fat-laden macrophages. There was no evidence of any inflammatory reaction.

The liver showed a diffuse fatty change. The hepatocytes were filled with numerous vacuoles

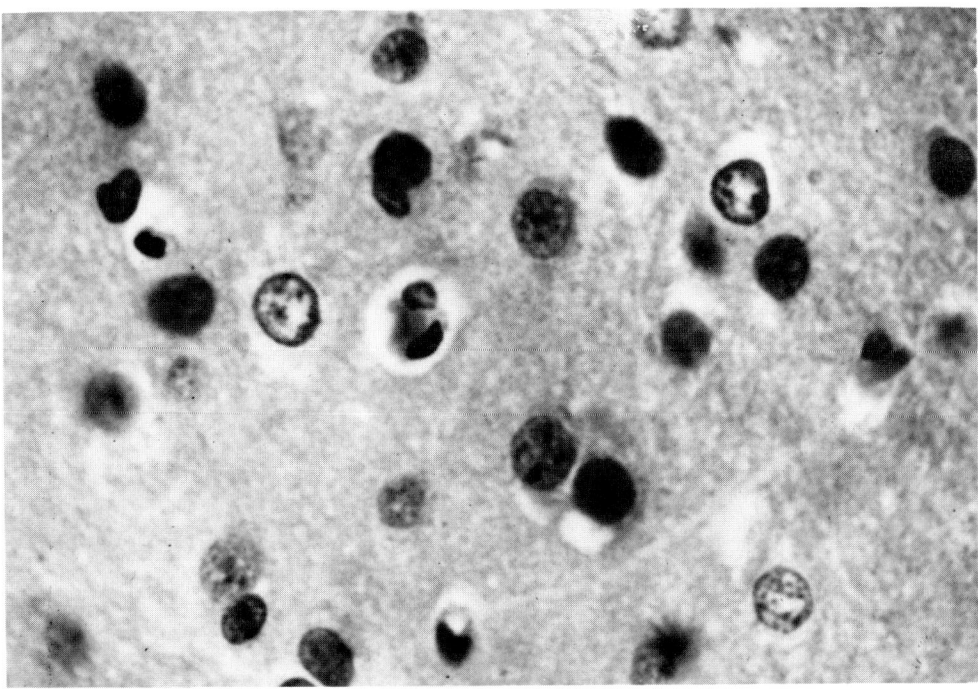

FIGURE 3. Swollen astrocytes with pale nuclei (hematoxylin and eosin, ×405).

which did not displace the nuclei and stained with Sudan in the frozen sections (Figure 4). The hepatocyte nuclei contained large nucleoli; the wavy nuclear membrane was occasionally disrupted. In two cases, eosinophilic intranuclear inclusions were present. The Kupffer cells were large and vacuolated; some had undergone eosinophilic necrosis manifested by karyorrhexis and pyknosis (Figure 5). Glycogen in the liver cells was substantially reduced or absent. Small foci of necrosis surrounded by mononuclear infiltrates with occasional polymorphs were found in five cases (Figure 6). Some of the lipid vacuoles contained birefringent crystals (Figure 7).

In the kidney, fatty degeneration was obvious in the proximal convoluted tubules and in the loops of Henle. Moving down the nephron, there were progressive decreases in the size and number of droplets (Figure 8).

Fibers of the myocardium showed steatosis, particularly pronounced in the subendocardial and subepicardial areas and in the bundle of His. (Figure 9). Granular decomposition and fatty degeneration of striated muscle fibers were commonly present (Figures 10 and 11).

Lymphoid hyperplasia was the most striking and consistent change seen in the alimentary tract. A microscopic examination revealed edema and depletion of lymphocytes in the lymphoid tissue. The germinal centers were enlarged, showing marked phagocytosis of lymphocytes and nuclear fragments. There was some degree of necrosis and hyalinization of the cells in the centers of the lymphoid follicles (Figures 12 and 13).

Focal lipid depletion and vacuolar degeneration of the zona fasciculata in the adrenal cortex were seen (Figure 14). The pulmonary alveoli and bronchi contained groups of macrophages, some of which contained fat, while others were necrotic (Figure 15).

C. BACTERIOLOGICAL, VIROLOGICAL, AND TOXICOLOGICAL FINDINGS

Bacteriological examinations (Table 9) of the bronchial contents, blood, cerebrospinal fluid, and intestinal contents did not reveal pathogens. Attempts to isolate varicella and herpetic

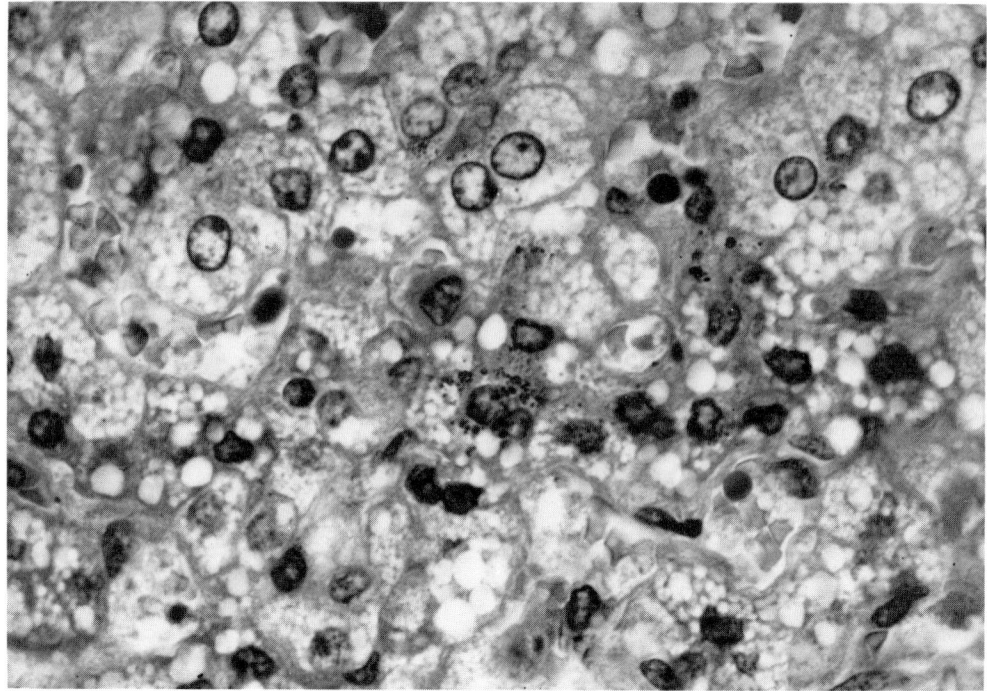

FIGURE 4. Diffuse fatty change of the liver. The hepatocytes are filled with fine vacuoles without displacement of the nuclei (hematoxylin and eosin, ×310).

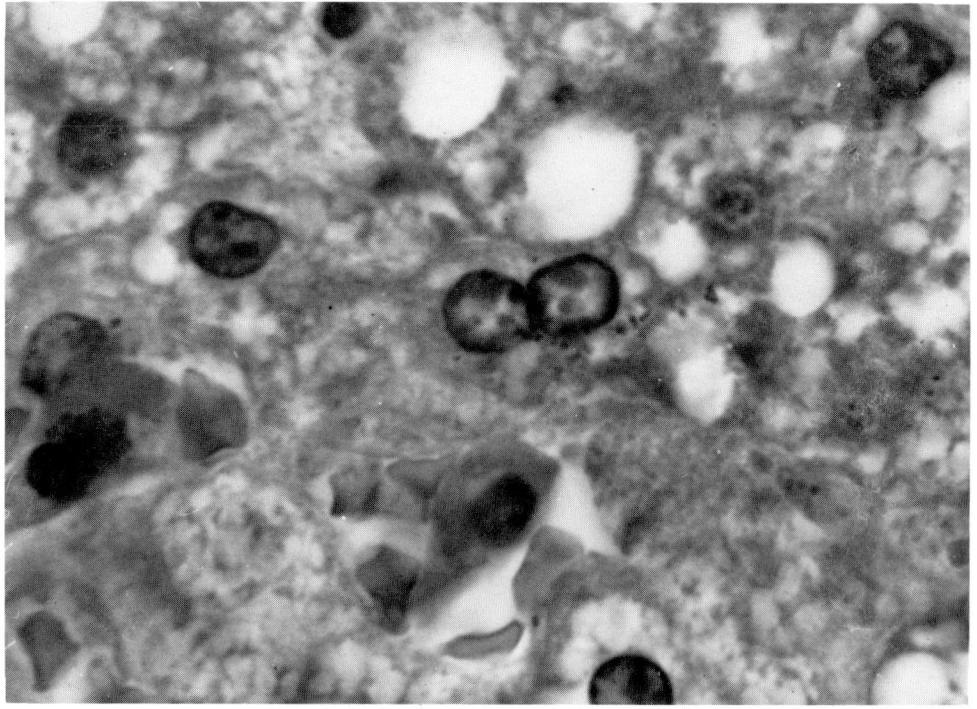

FIGURE 5. The swollen nuclei of the liver cells contain eosinophilic and amphophilic inclusions. Activation of the Kupffer cells (hematoxylin and eosin, ×810).

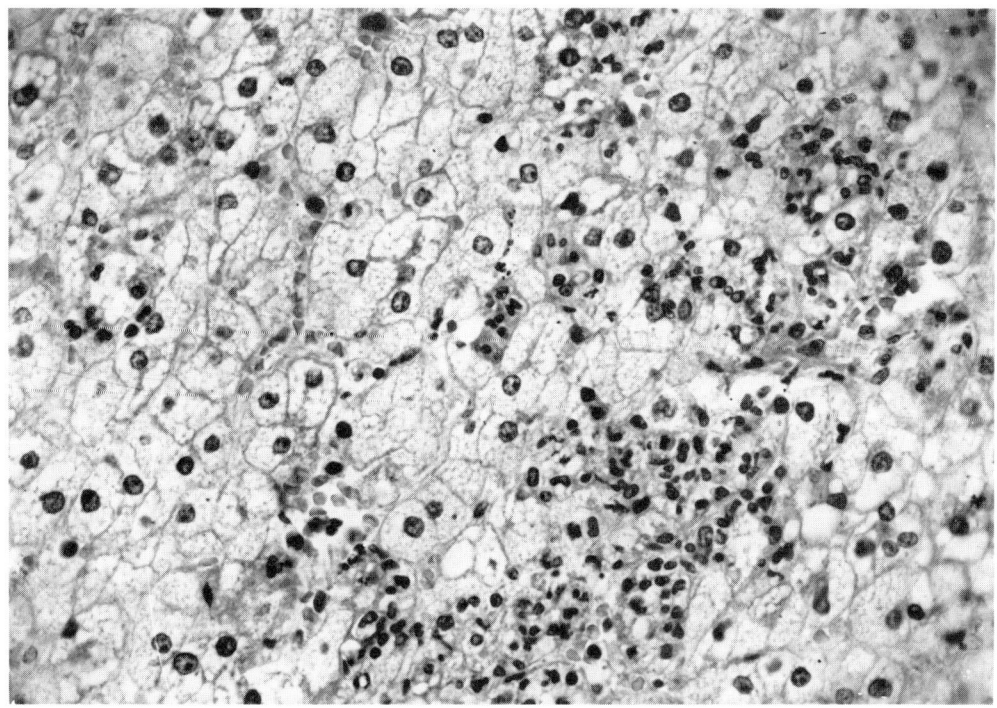

FIGURE 6. Inflammatory infiltration consisting of monocytes and polynuclear leukocytes in the liver (hematoxylin and eosin, ×265).

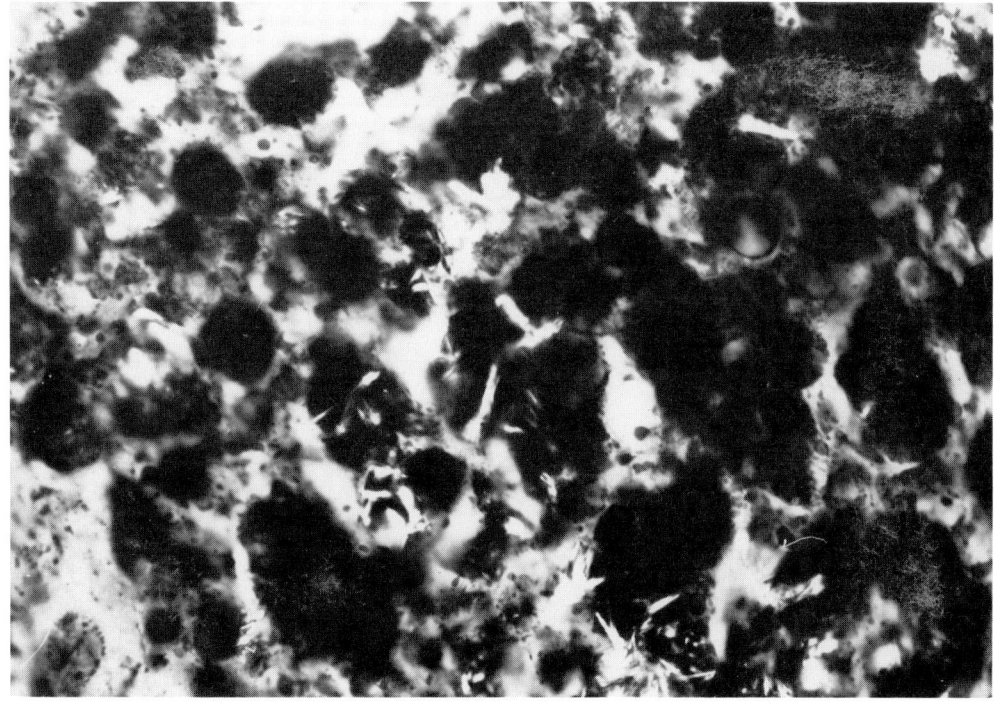

FIGURE 7. Birefringent irregular crystals in the fat vacuoles in the liver (Sudan, polarization ×132).

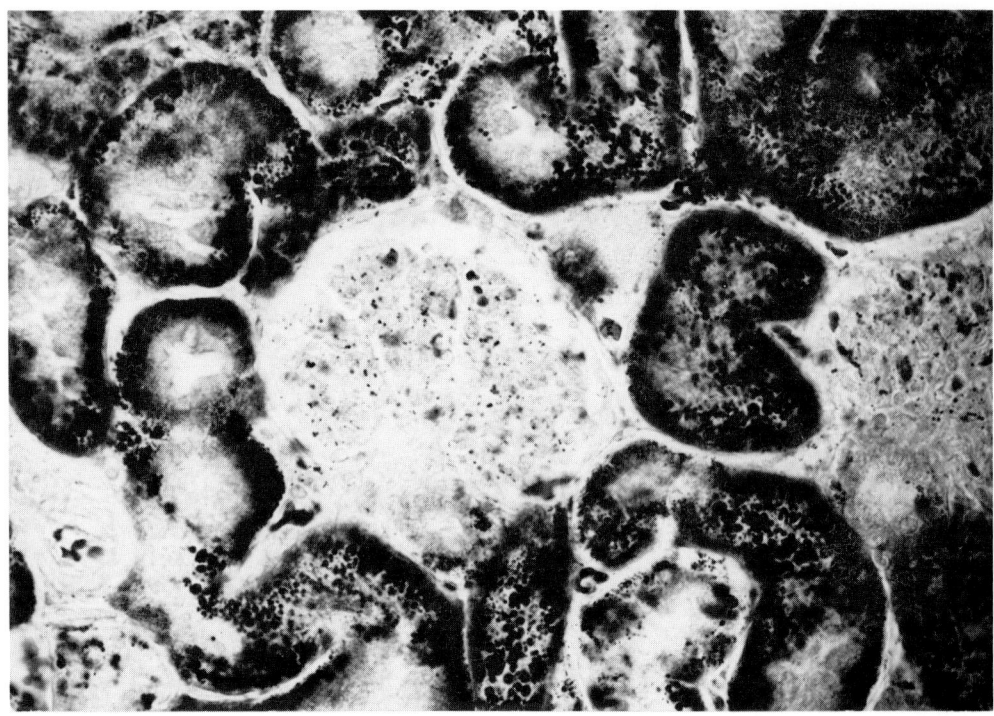

FIGURE 8. Fatty change of the primary convoluted tubules of the kidney (Sudan, ×132).

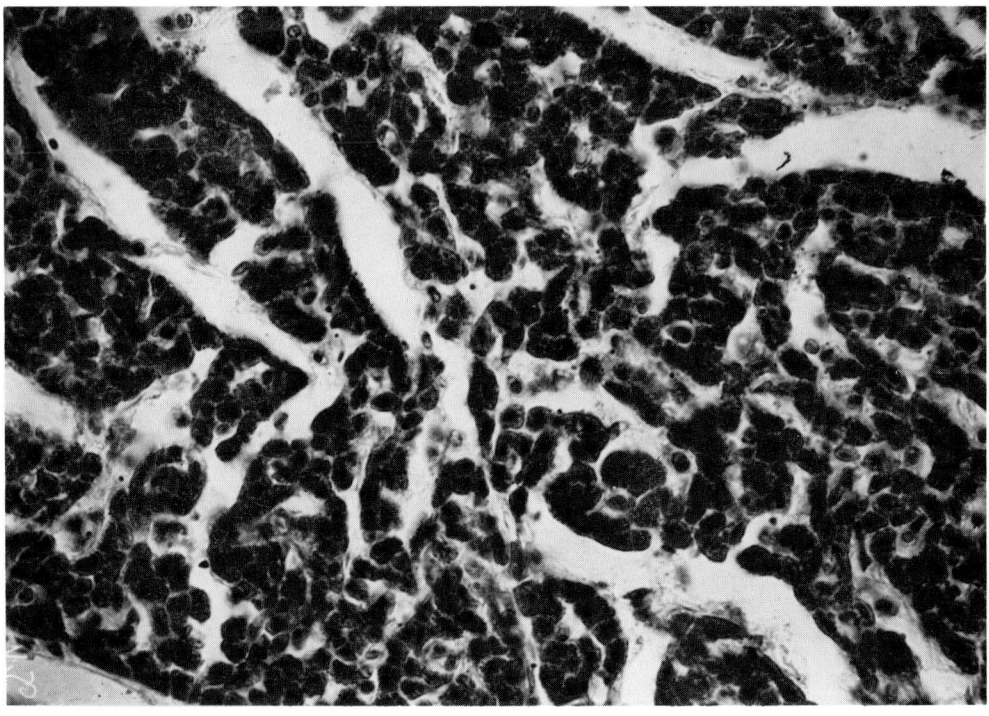

FIGURE 9. Fatty change of the myocardium (Sudan, ×132).

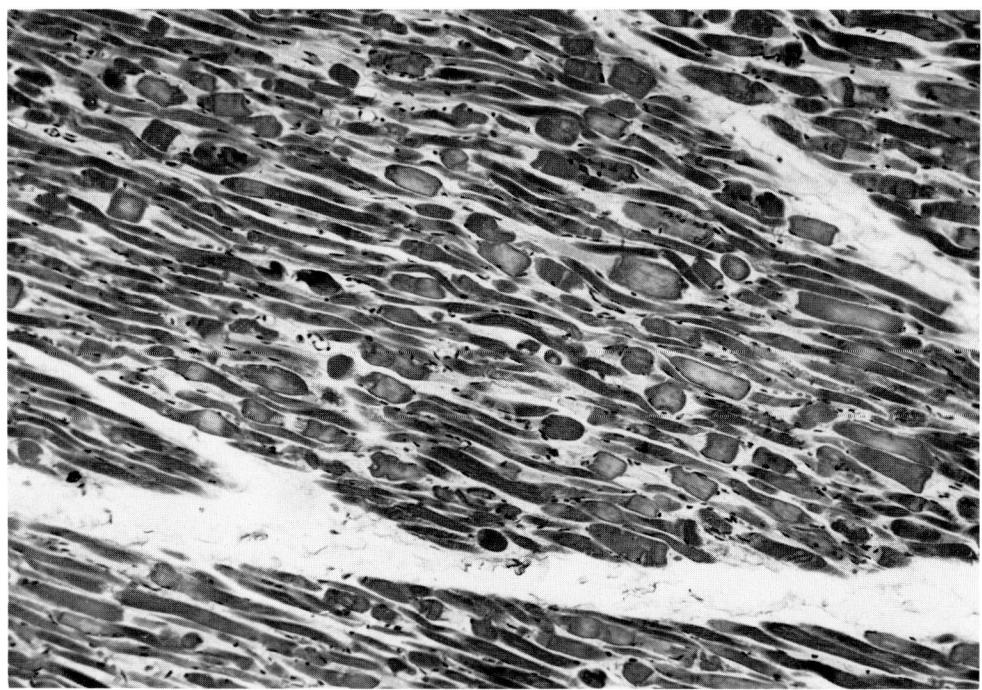

FIGURE 10. Granular decomposition of striated muscle fibers of *M. psoas* (hematoxylin and eosin, ×132).

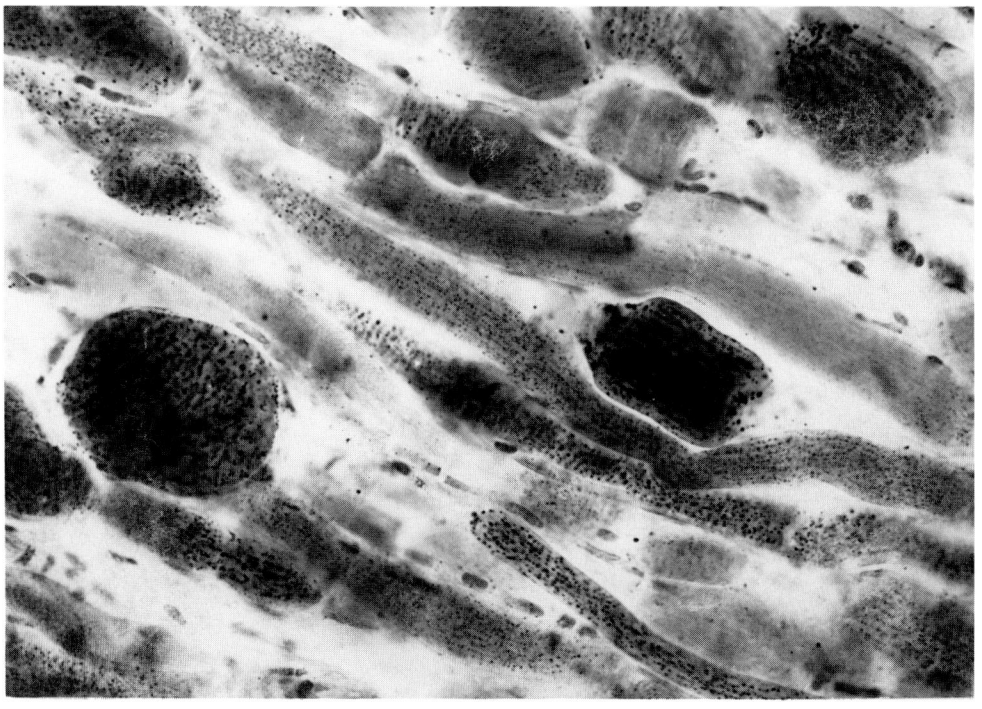

FIGURE 11. Fatty degeneration of muscle fibers of the diaphragm (Sudan, ×220).

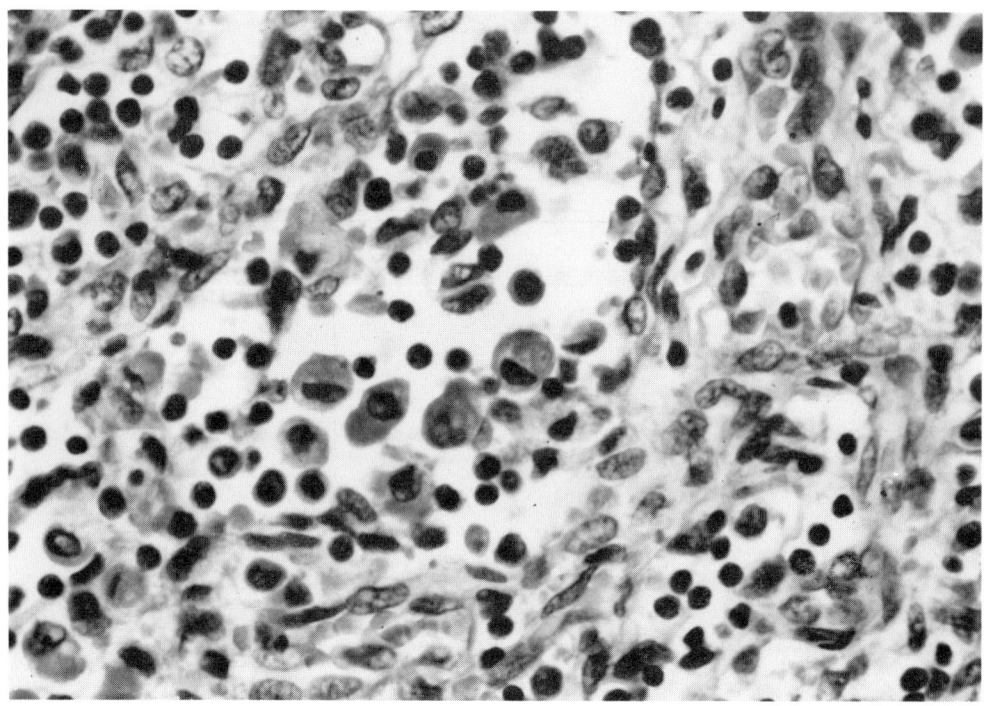

FIGURE 12. Swollen and necrotic histiocytes in the sinuses of the mesenteric lymph node (hematoxylin and eosin, ×440).

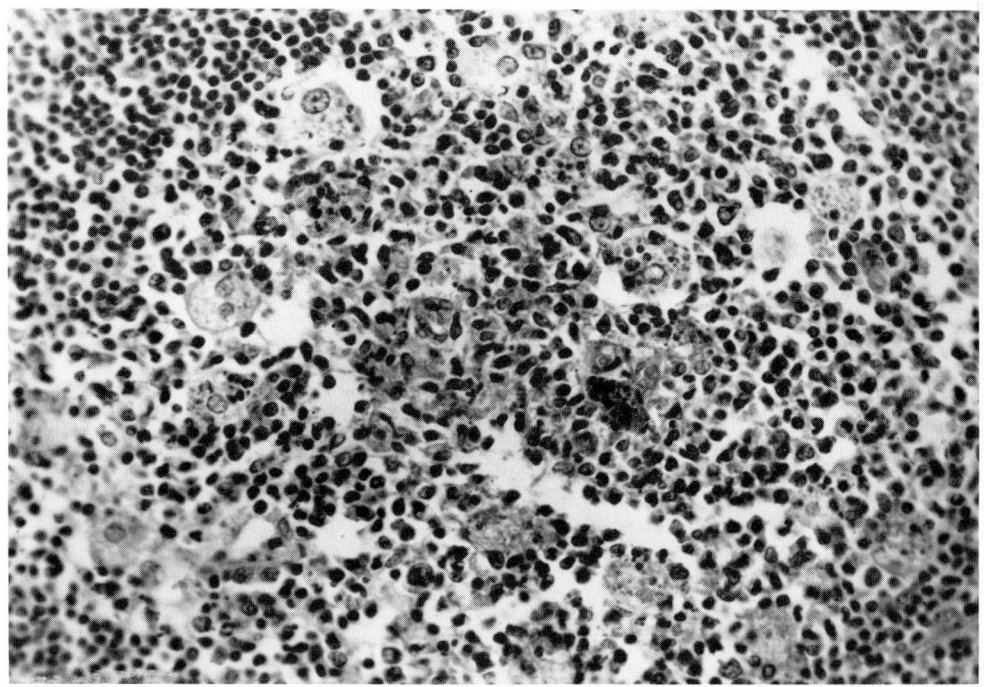

FIGURE 13. Phagocytosis of lymphocytes and nuclear fragments by reaction center macrophages in the spleen (hematoxylin and eosin, ×390).

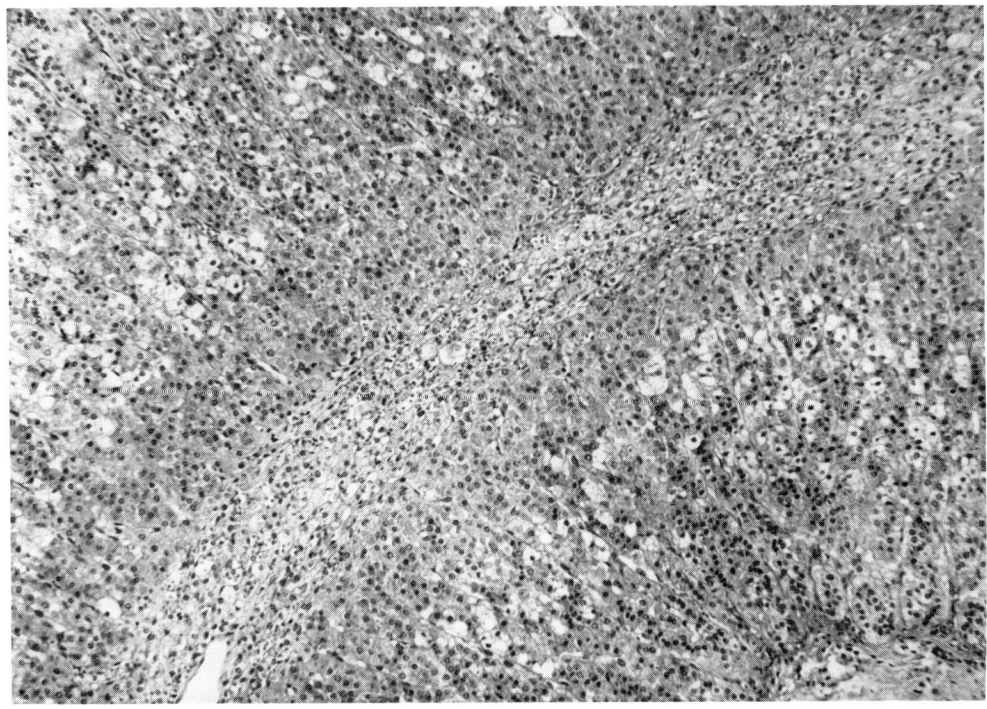

FIGURE 14. Vacuolar degeneration of the cells in the adrenal cortex (hematoxylin and eosin, ×132).

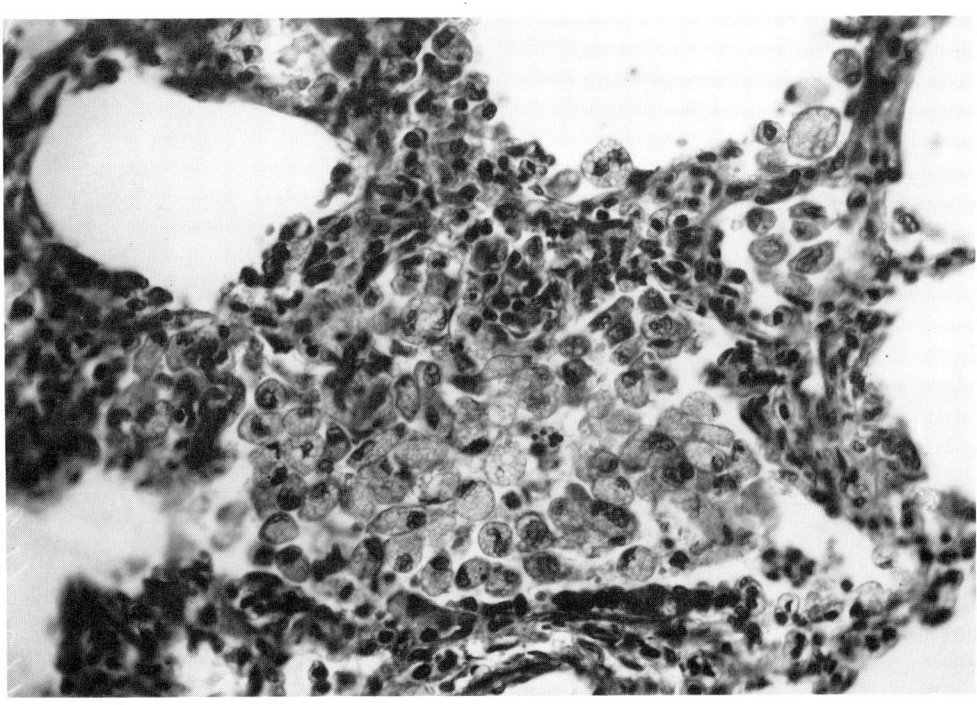

FIGURE 15. Groups of macrophages in the alveoli of the lung. Some macrophages contain fat, others are necrotic (hematoxylin and eosin, ×260).

TABLE 9
Virological and Toxicological Investigations in Reye's Cases (1958—1970)

Case no.	Influenza	Herpes	C-virus	Adeno	Varicella	Toxicological findings
1	–	–	–	–		–
2						–
3						
4						
5						
6	–			–		–
7		–	–			–
8	–			+		–
9				–		
10	–			–		
11	–			–		
12	–	–		–		
13	–	–		–		
14			–	–		
15				–	–	
16				–		
17	–			–		
18				–		
19	–			–		

viruses from the brain, influenza virus from the lung, laryngeal contents and serum, and C-virus from the heart and serum were unsuccessful. Out of 14 attempts to isolate adenovirus, only one was successful and adenovirus type 3 was isolated from the liver, tonsils, and conjunctiva.

Samples of organs (brain, heart, liver, kidney, stomach), gastric contents, stool, and urine of five cases studied by the laboratory for Toxicology and Forensic Chemistry in Prague did not reveal the presence of lead, arsenic, phosphorus, alkaloids, barbiturates, or histamines.

D. EPIDEMIOLOGICAL DATA

Epidemiological investigation in ten Reye's patients (Table 10) revealed contacts with infectious hepatitis which occurred simultaneously or subsequently in ten patients' residences or nursery schools. It was of a particular interest that 10 d after the death of a five-year-old girl, an epidemic of infectious hepatitis occurred in the patients' nursery school and that the physician who had treated the child also fell ill with hepatitis. In four cases, a febrile flu-like illness in the family was noticed. In two cases, contacts with varicella and, in one case, with morbilli were found. In three cases adenovirus infection was proved serologically in the children of patient's nursery schools. More often affected children came from rural areas (12) compared with those from urban areas (7).

III. A 15-YEAR STUDY OF REYE'S SYNDROME IN RELATIONSHIP WITH AFLATOXIN, 1972 TO 1986

A. MATERIAL

The studied group consisted of 118 cases including 30 cases of RS patients autopsied in Hradec Králové from 1972 to 1986. The material of 88 cases containing formalin-fixed samples of organs and fresh-frozen liver tissue, in some instances also the bile, for a chemical investigation for the presence of aflatoxin has been supplied by courtesy of pathologists from Prague, Most, Písek, Plzen, Brno, and Ostrava. The material has been completed with clinical information and gross pathological findings.

The liver samples of 40 children aged from 3 d to 10 years who died of unrelated diseases were

TABLE 10
Epidemiological Data in Reye's Cases (1958—1970)

Case no.	Place of residence		Contacts with viral infections in the families or nursery schools				
	Urban	Rural	Hepatitis	Influenza	Varicella	Morbilli	Adeno
1		+	+				
2	+		+				
3		+				+	
4		+	+				
5	+		+				
6	+		+				
7	+		+	+			
8		+	+				
9	+			+			
10		+	+				+
11		+					+
12	+			+	+		
13		+	+				+
14		+	+				
15		+			+		
16	+						
17		+					
18		+		+			
19		+					

used as controls. This group involved children with bronchopneumonia, respiratory distress syndrome, meningitis, gastroenteritis, congenital heart diseases, hemoblastosis, brain tumors, congenital bile-duct atresia, neonatal giant-cell hepatitis, and accident.

Samples of the food consumed by the children shortly before the onset of the disease have been collected in 17 households and investigated for the presence of aflatoxin.

B. METHOD

A routine histological examination has been performed in all cases. The sections were stained with hematoxylin and eosin, Sudan, Gomori, and PAS. The liver biopsy, as well as samples of the liver and kidney, taken in 3 cases within 2 h after death were fixed in 3% glutaraldehyde, post-fixed in 1% osmium tetroxide, dehydrated in ethanol, and embedded in Durcupan for electron microscopy.

The aflatoxin content of the tissue samples was determined in 117 cases by chromatography, spectrophotometry, and radioimmunoassay (RIA). In addition to chemical analyses, a biological test on 2-d-old ducklings was carried out.

C. CHEMICAL ANALYSIS
1. Thin Layer Chromatography (TLC) and Spectrophotometry

Chemicals — Authentic aflatoxin B_1 and G_1 (Calbiochem California) were used as standards; their UV absorption and fluorescence excitation and emission spectra at a concentration of 1.6 mg in 100 ml of chloroform are presented in Figures 16 and 17. Aflatoxin M_1 was supplied courtesy of Prof. Dr. F. Kiermeier (Technische University, München, FRG).

Preparation of samples — Homogenized liver samples (20 to 50 g) were subjected to lyophilization. The samples were extracted by chloroform. The evaporation residues free of chloroform were dissolved in a methanol-acetone mixture. Ten ml of 20% aqueous solution of lead acetate was added. The precipitate obtained was separated by centrifugation, and the solution was poured into the separatory funnel. The solution was three to five times extracted

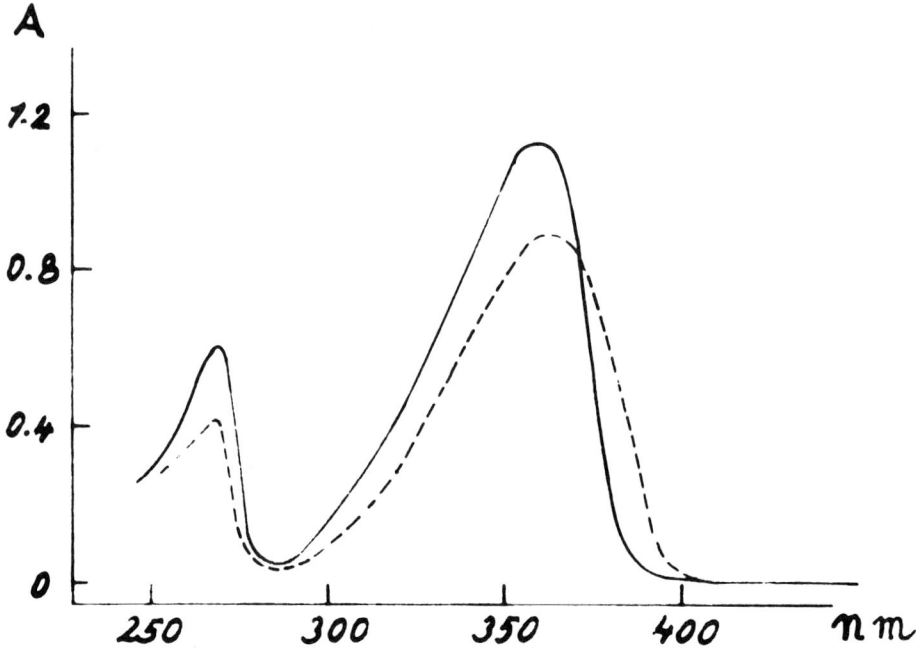

FIGURE 16. UV absorption spectra of aflatoxin B_1 (full line) and G_1 (broken line).

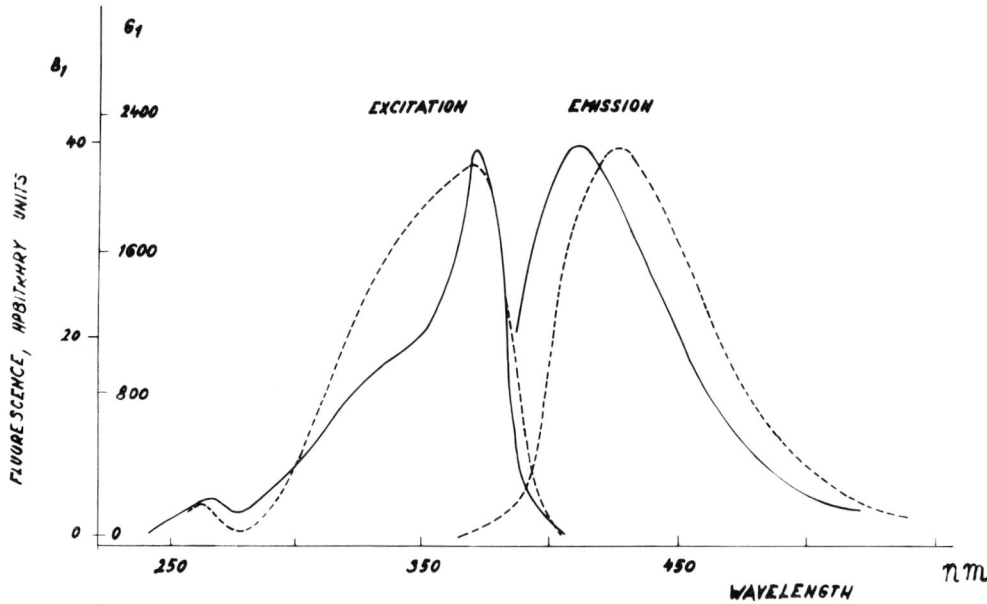

FIGURE 17. Excitation and emission spectra of aflatoxin B_1 (full line) and G_1 (broken line). Spectral maxima (in nanometers) from these measurements: aflatoxin B_1 (full line): absorption 265 and 359, fluorescence-excitation 266 and 374, emission 416; aflatoxin G_1 (broken line): absorption 265 and 359, fluorescence-excitation 263 and 372, emission 430.

by 50 ml n-hexane, and the aqueous layer was three times extracted by 50 ml chloroform. The chloroform extracts were pooled and filtered through anhydrous sodium sulphate, and chloroform was stripped into the rotary vacuum evaporator. The evaporated residues were dissolved in a small amount of chloroform and its small portion tested on the Kieselgel G (type 60) with aflatoxin standards in eluting systems as follows: chloroform-acetone (90 + 10), benzene-acetic acid-methanol (90 +5 + 5), chloroform-acetone-isopropanol (85 + 10 + 5), and chloroform-methanol (97 + 3). For spectrophotometric measuring, the substantial portion of the evaporation residue was put on the chromatographic column of Kieselgel H (type 60) and aflatoxin eluted by using the system chloroform-acetone (90 + 10). The solvent was removed from the aflatoxin containing fraction by evaporation, the residue having been dissolved in methanol and then measured spectrophotometrically. (The spectra were measured with SP 800 Unicam Spectrophotometer and D.C. 3000 Ciampolini Spectrofluorimeter, which operates by automatically controlled excitation energy). Food samples have been investigated according to the method of Stubblefield.[4]

Analysis — Analysis of the biological material was performed in two laboratories: The Chemical Laboratory of the State Veterinary Institute in Hradec Králové, by Dr. F. Brodsky, and the Laboratory of Experimental Medicine of the Czechoslovak Academy of Sciences in Prague, by D. Vesely. Six liver samples were assayed by two-dimensional chromatographic technique in the Mycotoxin Research Laboratory in Bilthoven, Netherlands, by Drs. Schuller and Egmond. Food samples were assayed in the Laboratory of Food Hygiene in Plzen by Drs. Turek and Adensam.

2. Radioimmunoassay (RIA)

The commercial RIA test for aflatoxin B_1 (The Institute of Radioecology, Kosice, Czechoslovakia) was used. Rabbit antisera were raised against aflatoxin B_1 (AFB_1), using a conjugate of AFB_1-O-carboxymethyloxime and bovine serum albumin. Antiserum exhibited a cross-reaction to $AFB_2 < 10\%$, $G_1 < 3\%$, $G_2 < 0.2\%$, $M_1 < 0.1\%$ and was highly specific to AFB_1.

Sample preparation — 1 g of liver tissue was homogenized in 10 ml of a saline solution, then extracted by 25 ml of chloroform. The solvent was evaporated in a test tube and the residue dissolved in 0.5 ml of 3% casein for 24 h at 37°C. A volume of 0.1 or 0.05 ml of this casein solution was used for RIA.

RIA procedure — A 5 ml centrifugation tube was supplied with 0.1 ml of the sample solution, 0.1 ml of the antibody solution, and 0.1 ml of the solution of (^{3}H)-aflatoxin, and the mixture was incubated for 24 h at +4°C. After incubation, the mixture was completed with 0.5 ml of the 0.3% suspension of activated NORIT charcoal and the sample was centrifuged. Counting vials containing 5 ml of the scintillation liquid (SLD 41, Spolana, Neratovice, Czechoslovakia) were supplied by 0.5 ml of the supernatant, and radioactivity was measured on a Prias-Packard Instrument. The results were read from a calibration curve summarized from data obtained in an analogous manner from the solutions of standard AFB_1 of appropriate dilutions (Figure 18).

The RIA procedure has been carried out in the Isotope Laboratory of the Czechoslovak Academy of Sciences in Prague, by Drs. K. Veres and D. Píchova.

D. CLINICAL AND LABORATORY FINDINGS

The studied series (Table 11) consisted of 118 children: 63 boys and 55 girls. Seventy-two of them were children under 1 year of age (Figure 19). The disease occurred throughout the year peaking in April and September (Figure 20). The highest number of cases occurred in 1976 and 1978. Since 1981, a decreasing tendency of the disease has been noticed (Figure 21). In 96 cases (81%), the disease started with upper respiratory infection. Varicella was recognized as the preceding illness in three patients. In five cases, the disease began with gastrointestinal symptoms. Eleven children were vaccinated 5 to 7 d prior to the onset of the prodromal symptoms; four were inoculated with a smallpox vaccine, two were given a morbilli vaccine, and five received a vaccine of live oral polio virus. Salicylate therapy was used in 23 out of 77 children at the onset of the disease. The other children were given antibiotics and corticoids.

Elevated levels of serum transaminases were found in 96 of 101 children in whom the

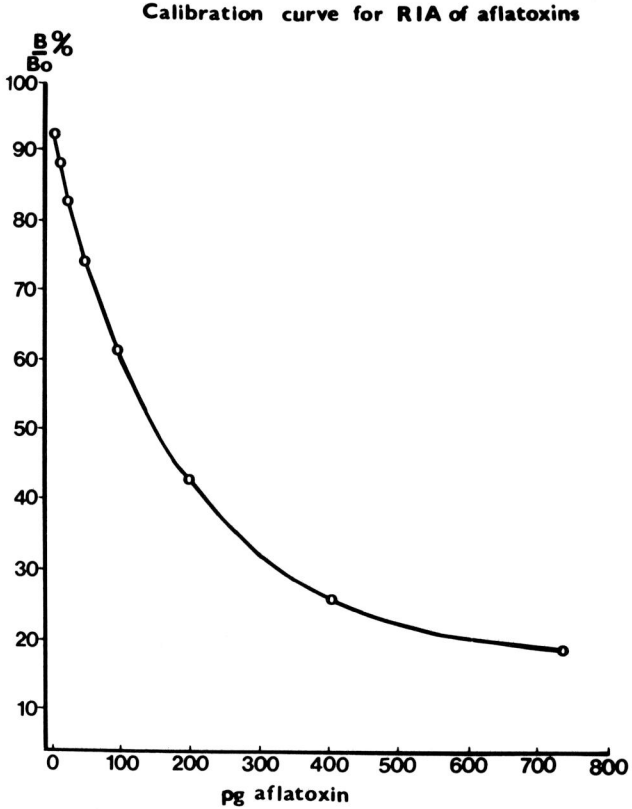

FIGURE 18. Calibration curve for RIA of aflatoxin.

TABLE 11
Clinical Data and Laboratory Findings in Reye's Cases 1972—1986

Antecedent illnesses	
URI	96
GIT disorder	5
Varicella	3
Vaccination	4 Smallpox
	2 Morbilli
	5 Polio
Elevated transaminases	96 (101)
Hyperammonemia	39 (51)
Hypoglycemia	46 (73)
Prolonged prothrombin time	93 (105)
Salicylate therapy	23 (77)
Death within 5—10 d	112

Note: Numbers in brackets = number for whom information recorded.

measurements were performed. The blood ammonia level was investigated in 51 patients and found elevated in 39 cases. Low blood sugar level was present in 46 out of 73 investigated cases. A prolonged prothrombin time was found in 93 out of 105 patients.

In all but six cases, the course of the disease was acute and resulted in death within 5 to 10 d after the first signs had appeared.

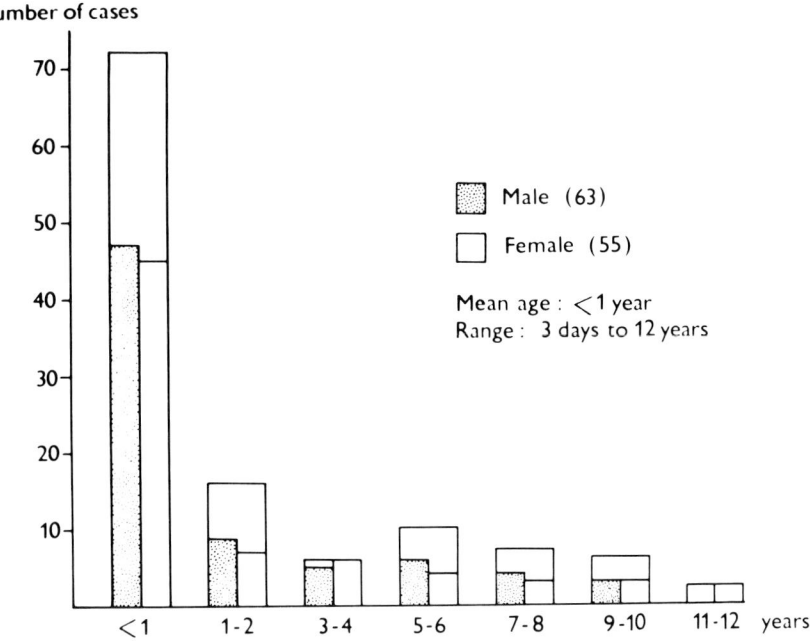

FIGURE 19. Reye's syndrome cases 1972 to 1986. Age and sex distribution.

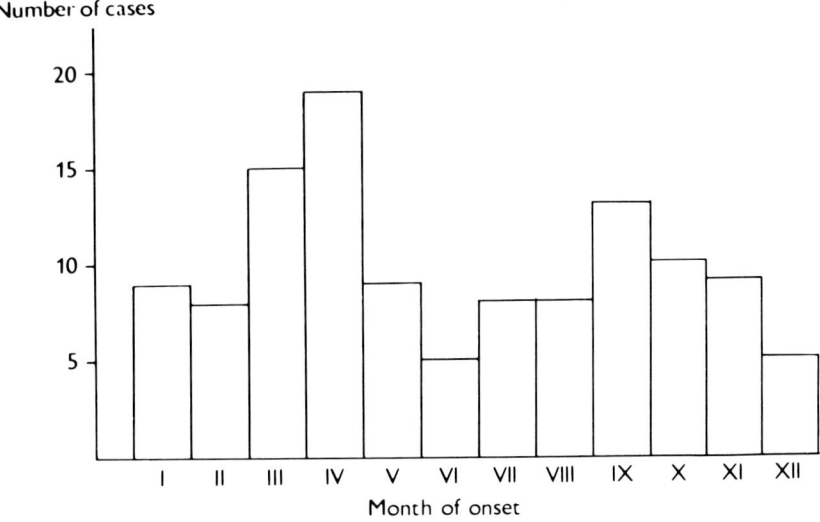

FIGURE 20. Reye's syndrome cases 1972 to 1986. Month of onset of 118 reported cases.

E. EPIDEMIOLOGICAL DATA

An epidemiological investigation (Table 12) revealed that the disease occurred more often in children from rural regions (73) compared with those from urban areas (45). Twenty-one attempts to isolate the influenza virus gave negative results. Out of 30 attempts to isolate adenovirus, only 4 were successful, and adenovirus type 3, 5 and, in two cases, adenovirus type 2 were recovered from the liver and mesenteric lymph nodes. Serological investigations revealed positive influenza A antibodies in 7 out of 13 investigated cases, and, in two cases, positive hepatitis A antibodies were found.

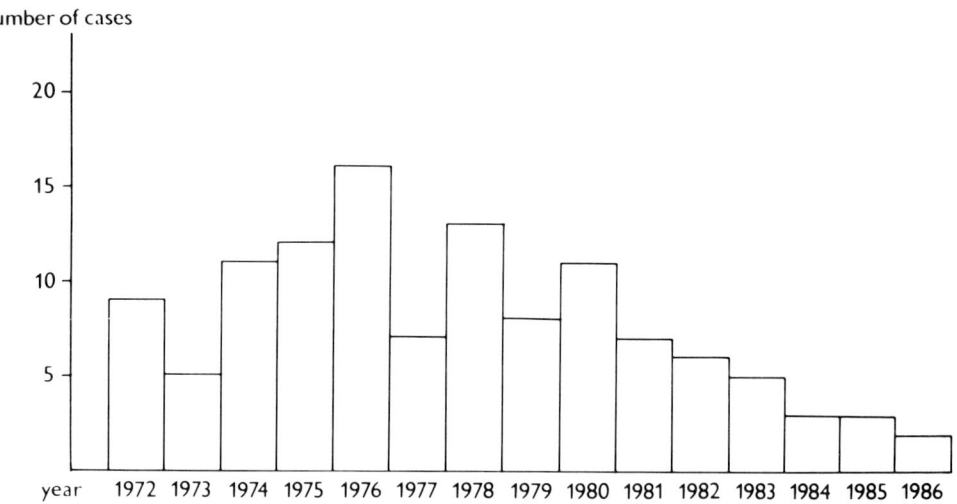

FIGURE 21. Reye's syndrome cases 1972 to 1986. Occurrence of 118 reported cases in individual years.

TABLE 12
Epidemiological and Virological Findings in Reye's Cases 1972—1986

Place of residence		Contacts with viral infections		Virus isolation	Serology	
Urban	Rural					
45	73	Flu-like	13	Influenza	21 Negative (21)	Influenza A
		Adeno	5	Adeno	4 Positive (30)	7 (13)
		Hepatitis	2			Hepatitis A
		Varicella	2			2 (5)
		Not known	96			

Note: Numbers in brackets = numbers for whom information available.

F. AFLATOXIN IN REYE'S SYNDROME CASES

The liver extracts of patients with the presence of aflatoxin showed spots with blue fluorescence like that of AFB_1 in 365 nm UV light, the same color changes as AFB_1 when treated with 50% H_2SO_4, and R_F values identical to those of the commercial sample of AFB_1 (Figure 22). The absorption spectra of aflatoxin isolated from the patients[1] livers and food samples were identical to those of standard AFB_1, as Figure 23 indicates. The aflatoxin levels of individual Reye's cases are presented in Tables 13 to 27.

Out of 117 liver samples and 10 bile samples investigated for the presence of aflatoxin, aflatoxin B_1 has been proven in 71 (Figure 24). Concentrations of aflatoxin varied from 5 to 1760 µg/kg in the liver, with an average between 80 to 200 µg/kg. Liver samples with the highest concentrations were sent to the Mycotoxin Research Laboratory in Bilthoven to check our results. The concentrations found in that laboratory did not differ significantly from those found in our laboratories.

In addition to AFB_1, aflatoxin M_1 at concentrations of 0.8 to 20 µg/kg in four liver samples and two undefined metabolites of the aflatoxin type in the liver of a 3-d-old boy were found. Out of 40 control liver samples of children who died of unrelated diseases, only in 3 cases (2 meningitis, 1 enterocolitis) has less than 2 µg/kg of AFB_1 been detected.

AFB_1 was found in 12 out of 17 samples of powdered milk and in one packet of meal collected in the households of the patients. The concentrations ranged from 250 to 5400 µg/kg. In one

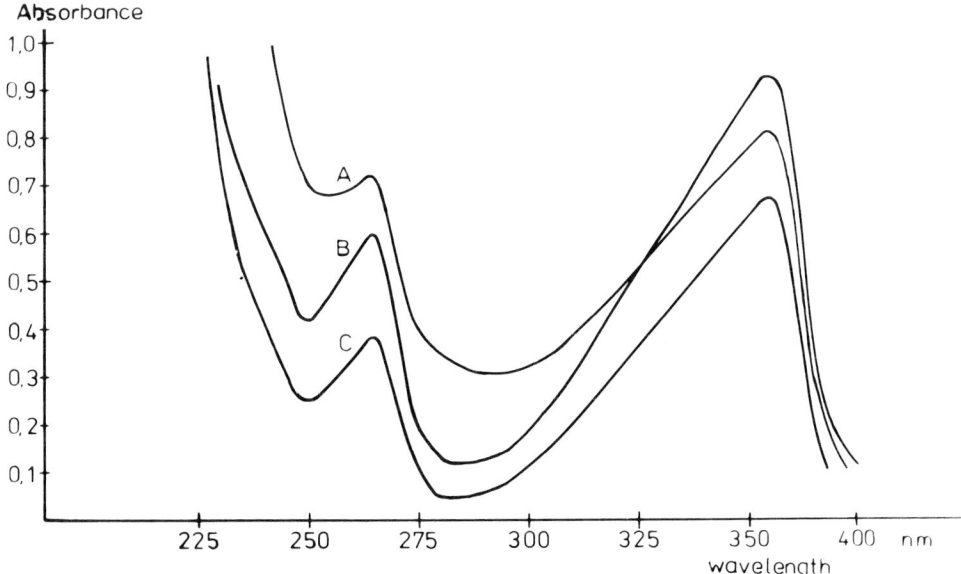

FIGURE 22. Chromatogram of the liver specimens and standard of aflatoxin B_1 and G_1: (1) authentic aflatoxin B_1 (upper) and G_1 (lower), (2) liver extract of the patient with added authentic aflatoxin B_1 and G_1, and (3) liver extract of the patient without addition.

sample, traces of AFB_2 and AFG_2 have been proven. As expected, aflatoxin M_1 has not been detected in milk. Therefore, it seems possible that aflatoxin contamination was rather secondary, occurring either during the processing or storage.

G. BIOLOGICAL TEST

Nine 2-d-old ducklings were given a single dose of 50 µg of the substance extracted from the liver of one patient. Two of the ducklings died spontaneously within 48 h, the others were killed 10 d and 4 weeks later. Clinical symptoms of the ducklings included feebleness, convulsions, pareses, and hemorrhagic diathesis. The livers of the ducklings which died spontaneously during the first 2 d showed a diffuse fatty change, focal cell necrosis, and hemorrhages (Figure 25). The livers of the ducklings killed after 10 d revealed steatosis and periportal fibrosis with a severe bile-duct proliferation (Figure 26). After 4 weeks, the liver lesions included a fatty change and cirrhosis with the presence of hyperplastic nodules composed of basophilic hepatocytes (Figures 27 and 28).

H. MORPHOLOGICAL FINDINGS

Autopsy findings as well as histological changes in the patients of this series did not differ from the findings on the group of 19 children with RS which were described earlier.

The ultrastructural study was performed on a liver biopsy sample taken from a 7-month-old boy at the comatose stage (Table 16, case 3) and on the liver and kidney specimens taken within 2 h after death from a $4\frac{1}{2}$-month-old girl and a 5-year-old boy (Table 15, cases 6 and 8). Aflatoxin B_1 has been chemically proved in the liver samples of all three patients.

The ultrastructural changes of the liver biopsy sample revealed many fat droplets and glycogen depletion in the cytoplasm of the liver cells (Figure 29). The mitochondria were swollen and pleomorphic, with matrix rarefaction and a loss of dense bodies. The cristae were reduced in number and occasionally their fragmentation was found (Figure 30). The nuclei were swollen and islands of condensed chromatin were found along the inner surface of the nuclear membrane. Clusters of granular material, identical to interchromatine granules, were present in

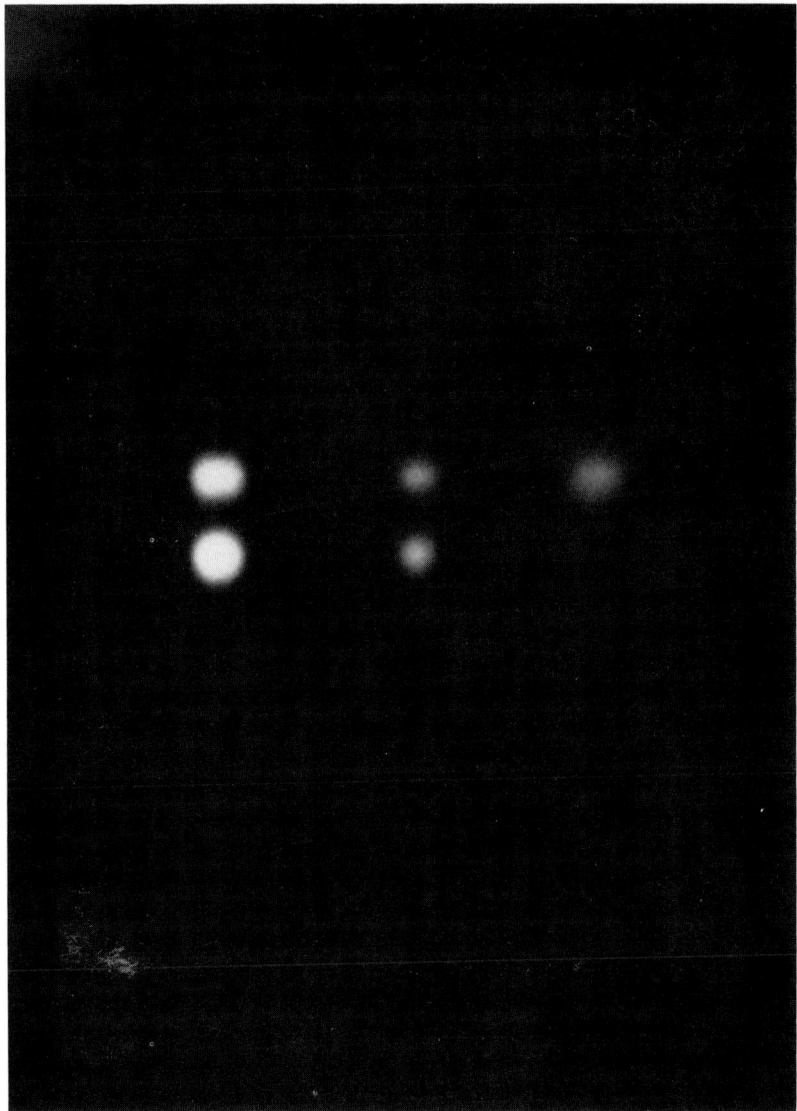

FIGURE 23. Absorption spectra of aflatoxin B_1 from liver sample, food sample, and standard of aflatoxin B_1: curve A — aflatoxin B_1 isolated from the liver; curve B — aflatoxin B_1 isolated from powdered milk; and curve C — standard of aflatoxin B_1.

the nucleoplasm (Figure 31). The nucleoli were enlarged, compact, and some exhibited a dense zone and a zone of moderate density containing vacuoles with dense macrogranules. Nucleolar bodies were seen to lie adjacently to the nucleolus (Figure 32).

Identical ultrastructural alterations were found in the liver and renal tubular cells of the specimens taken within 2 h after the death of the other two children (Figures 33 to 37).

The mitochondrial ultrastructural alterations corresponded to those which are considered as highly characteristic of RS, while the nuclear abnormalities found in these Reye's cases were quite uncommon.

Simard and Bernhard[5] in their study of the effects of various antimetabolites observed that segregation and zoning of granular and fibrillar nucleolar structures is a characteristic alteration produced by substances that bind with DNA and interfere with its template activity in DNA-directed RNA synthesis.

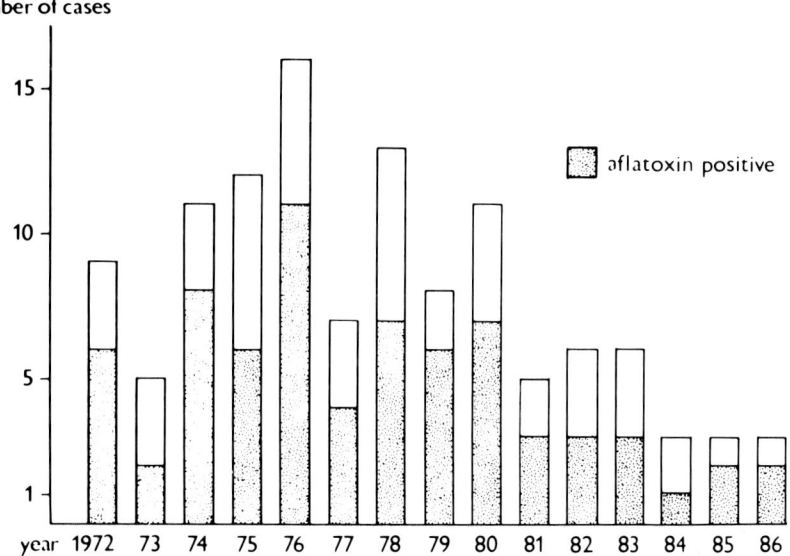

FIGURE 24. Aflatoxin detected in Reye's syndrome patients (1972 to 1986).

TABLE 13
Aflatoxin in Reye's Syndrome Cases 1972

	No.	Age[a]	Sex	AFB$_1$ (µg/kg) Liver	Food
1.	J.D.	22 m	F	120	—
2.	I.K.	8 m	F	80	—
3.	M.K.	2 yr	M	Negative	—
4.	J.V.	3 d	M	80	—
5.	V.M.	7 m	M	102	—
6.	V.Č.	8 m	M	70	—
7.	L.D.	3 m	F	35	—
8.	L.C.	3 m	M	Negative	—
9.	P.P.	3 m	M	Negative	—

[a] yr = years, m = months, and d = days.

TABLE 14
Aflatoxin in Reye's Syndrome Cases 1973

	No.	Age[a]	Sex	AFB$_1$ (µg/kg) Liver	Food
1.	J.U.	3.5 yr	F	Negative	—
2.	J.H.	7 m	M	Negative	—
3.	J.K.	6 m	M	20	—
4.	A.H.	4 yr	M	Negative	—
5.	H.A.	4 m	F	80	—

[a] yr = years, m = months.

TABLE 15
Aflatoxin in Reye's Syndrome Cases 1974

No.		Age[a]	Sex	AFB$_1$ (µg/kg)	
				Liver	Food
1.	A.D.	16 m	F	120	—
2.	V.M.	1.5 m	F	70	—
3.	M.S.	3 m	F	20	—
4.	M.J.	14 m	M	1126 (1600)[b]	—
5.	L.P.	8 yr	M	80	—
6.	A.S.	22 m	F	230	590
7.	J.K.	4 m	M	102	—
8.	V.V.	8 yr	F	20	—
9.	J.T.	11 m	M	Negative	—
10.	P.J.	6 m	M	Negative	—
11.	M.S.	2 m	F	Negative	—

[a] yr = years, m = months.
[b] Investigated in Bilthoven.

TABLE 16
Aflatoxin in Reye's Syndrome Cases 1975

No.		Age[a]	Sex	AFB$_1$ (µg/kg)	
				Liver	Food
1.	L.V.	4 m	M	1760 bile 33	520
2.	D.P.	15 m	F	Negative	—
3.	L.H.	15 m	F	60	—
4.	Z.C.	5 m	M	450	—
5.	J.K.	10 m	M	Negative	—
6.	T.V.	4 m	M	513 bile 320	320
7.	K.B.	16 m	M	Negative	—
8.	M.Š.	4 m	M	194 20 M$_1$	—
9.	J.H.	8 yr	M	Negative	—
10.	M.Ch.	3 m	M	580	—
11.	M.R.	7 m	M	Negative	—
12.	P.H.	4 m	F	Negative	—

[a] yr = years, m = months.

Nucleolar zoning has been reported in rat and monkey liver cells following acute administration of AFB$_1$.[6,7] Since AFB$_1$ was demonstrated to bind covalently to DNA and to exhibit RNA and protein synthesis, it has been suggested that AFB$_1$ is responsible for the nucleolar abnormalities.

Bauer and collaborators[8] have reported nucleolar alterations in the liver cells of aquarium fish *(Brachydanio rerio)* and in the tentacle epithelial cells of a molusk *(Planorbarius corneus)* produced by exposure to AFB$_1$. Two zones of different densities were usually apparent in the nucleoli. Of special interest were vacuoles containing clusters of opaque granules, markedly similar in size and shape. An increase in interchromatine granules with AFB$_1$ at the acute stage

TABLE 17
Aflatoxin in Reye's Syndrome Cases 1976

				AFB$_1$ (µg/kg)	
	No.	Age[a]	Sex	Liver	Food
1.	M.R.	7.5 m	M	1140	5400
2.	M.P.	6 m	M	Negative	—
3.	R.B.	3 m	M	20	—
4.	K.K.	3 m	M	120	—
5.	R.M.	6 m	F	172	—
					2 M$_1$
6.	M.Č.	5 yr	F	150	—
7.	O.V.	3 d	M	1015	—
					9 M$_1$
					450 X$_1$X$_2$
8.	R.T.	6 m	F	1134 (1400)[b]	—
					0.8 M$_1$
9.	R.Z.	10 m	M	Negative	—
10.	J.M.	4 m	M	10	Negative
11.	I.R.	5 m	F	270	—
12.	M.H.	5 m	M	170	—
13.	K.P.	10 m	F	42	—
14.	M.G.	9 m	F	Negative	—
15.	J.K.	8 yr	M	Negative	—
16.	K.J.	11 yr	F	Negative	—

[a] yr = years, m = months, and d = days.
[b] Investigated in Bilthoven.

TABLE 18
Aflatoxin in Reye's Syndrome Cases 1977

				AFB$_1$ (µg/kg)	
	No.	Age[a]	Sex	Liver	Food
1.	J.K.	4 m	M	700	—
2.	V.M.	7 yr	M	250	—
3.	M.J.	3 m	F	81	1220
4.	E.B.	3 m	F	135	—
5.	P.G.	2.5 yr	M	Negative	—
6.	Š.P.	11 yr	F	Negative	—
7.	L.S.	10 yr	—	Negative	—

[a] yr = years, m = months.

has been reported in the rat liver.[9] This finding was considered interesting since only rarely was there some change reported in these granules in experimental conditions. It has been assumed that the increase in interchromatine granules may represent a compensatory synthesis of RNA at the extranucleolar sites, since aflatoxin, like many other carcinogens, inhibits nucleolar RNA synthesis in acute experiments.

The similarity between the ultrastructural alterations, particularly those of the nucleoli found in Reye's cases and those reported in animals treated with AFB$_1$, supports the hypothesis that aflatoxin may be involved in the etiology of some cases of RS.

TABLE 19
Aflatoxin in Reye's Syndrome Cases 1978

No.		Age[a]	Sex	AFB$_1$ (µg/kg)	
				Liver	Food
1.	J.H.	6 m	M	1150 (1250)[b]	—
2.	J.G.	6 m	M	40 (64)[b]	—
3.	I.S.	5 yr	F	194 (204)[b]	—
4.	A.M.	7 m	M	20 (15)[b]	—
5.	M.N.	10 m	M	Negative	—
6.	E.K.	18 m	F	Negative	—
7.	E.S.	6 m	F	Negative	—
8.	J.K.	6 yr	F	95	—
9.	J.H.	3 m	M	Negative	—
10.	Š.P.	10 yr	F	145	—
11.	M.V.	3 m	M	Negative	—
12.	S.H.	3 yr	F	48	—
13.	Š.Č.	8 m	F	Negative	—

[a] yr = years, m = months.
[b] Investigated in Bilthoven.

TABLE 20
Aflatoxin in Reye's Syndrome Cases 1979

No.		Age[a]	Sex	AFB$_1$ (µg/kg)	
				Liver	Food
1.	L.Č.	3 m	F	159	—
2.	I.H.	2 m	F	79	—
				bile 97	—
3.	M.Z.	6 yr	F	33	—
				bile 56	—
4.	R.P.	4.5 yr	M	75	—
				bile 86	—
5.	K.V.	10 m	M	Negative	—
6.	J.M.	9 m	M	280	3100
7.	D.B.	16 m	F	110	250
8.	J.B.	9 yr	F	Negative	—

[a] yr = years, m = months.

IV. REYE'S SYNDROME WITH CHRONIC LIVER DAMAGE

In six children of this series of 118 cases of RS, in three boys and three girls at the age of 3 to 14 months, the disease had a protracted course and ended in death within 2 to 4 months.

As usual the disease developed in them suddenly with febrile respiratory symptoms followed within 3 to 5 d by vomiting and with cerebral symptoms characterized by somnolescence and convulsions. Laboratory findings were leukocytosis, elevated serum transaminases, a prolonged prothrombin time, and, in two children, a low blood sugar level was present. Investigation of the cerebrospinal fluid and urine were negative. Liver biopsy was performed in an eight-month-old boy on the 3rd day of the illness, and the histological examination revealed a diffuse fatty change without any inflammatory reaction or necrosis (Figure 38).

After a short transient improvement, pareses and recurrent generalized convulsions ap-

TABLE 21
Aflatoxin in Reye's Syndrome Cases 1980

No.		Age[a]	Sex	AFB$_1$ (µg/kg)	
				Liver	Food
1.	A.U.	8 yr	M	94.6	—
2.	P.V.	7 yr	F	78	—
3.	S.D.	4 yr	F	94	—
4.	P.S.	8 m	M	Negative	—
5.	M.B.	9 m	F	Negative	—
6.	P.N.[b]	4.5 m	F	169	Negative
7.	V.H.	10 yr	F	Negative	—
8.	P.D.[b]	5 yr	M	130	450
9.	S.N.	6 yr	M	96	—
10.	J.H.	8 m	F	Negative	—
11.	P.N.	3 m	F	37	—

[a] yr = years, m = months.
[b] Liver and kidney samples taken for EMI.

TABLE 22
Aflatoxin in Reye's Syndrome Cases 1981

No.		Age[a]	Sex	AFB$_1$ (µg/kg)	
				Liver	Food
1.	J.M.	14 m	F	52	—
2.	J.S.	6 yr	F	Negative	—
3.	J.V.[b]	7 m	M	130	320
4.	D.J.	5 yr	M	36	—
5.	T.L.	6 yr	M	Negative	—

[a] yr = years, m = months.
[b] Liver biopsy taken for EMI.

TABLE 23
Aflatoxin in Reye's Syndrome Cases 1982

No.		Age[a]	Sex	AFB$_1$ (µg/kg)	
				Liver	Food
1.	P.B.	11 m	F	68	—
2.	J.V.	22 m	M	Negative	—
3.	V.Ch.	10 yr	M	Negative	—
4.	M.M.	4 m	F	125	—
5.	M.H.	7 m	M	183	570
6.	M.T.	8 yr	F	Negative	—

[a] yr = years, m = months.

peared, which dominated the clinical picture till death. Serum transaminases remained elevated. Investigations for inborn errors of metabolism as well as the genetic investigations of the parents and siblings gave negative results. The clinical diagnosis was a "subacute necrotizing encephalitis and encephalopathy with hepatopathy". The main clinical and laboratory findings are demonstrated in Table 28.

TABLE 24
Aflatoxin in Reye's Syndrome Cases 1983

No.		Age[a]	Sex	AFB$_1$ (µg/kg)	
				Liver	Food
1.	M.R.	6 m	F	159	610
2.	M.V.	13 m	M	Negative	—
3.	P.J.	5 m	M	Negative	—
4.	M.P.	2 yr	F	Negative	—
5.	M.B.	8 m	M	57	Negative
6.	M.K.	7 m	M	72.1	—

[a] yr = years, m = months.

TABLE 25
Aflatoxin in Reye's Syndrome Cases 1984

No.		Age (months)	Sex	AFB$_1$ (µg/kg)	
				Liver	Food
1.	L.H.	9 m	F	Negative	—
2.	M.V.	13 m	F	Negative	—
3.	I.K.	20 m	M	37.6	Negative

TABLE 26
Aflatoxin in Reye's Syndrome Cases 1985

No.		Age (months)[a]	Sex	AFB$_1$ (µg/kg)	
				Liver	Food
1.	R.L.	9 m	M	71.6 bile 60.5	560
2.	H.T.	3 m	M	67.8	—
3.	J.M.	18 m	M	Negative	—

TABLE 27
Aflatoxin in Reye's Syndrome Cases 1986

No.		Age[a]	Sex	AFB$_1$ (µg/kg)	
				Liver	Food
1.	K.K.	3 yr	F	Negative	—
2.	J.V.	2 m	M	80.1 bile 40.1	—
3.	P.K.	6 m	M	24.9 bile 15.8	—

[a] yr = years, m = months.

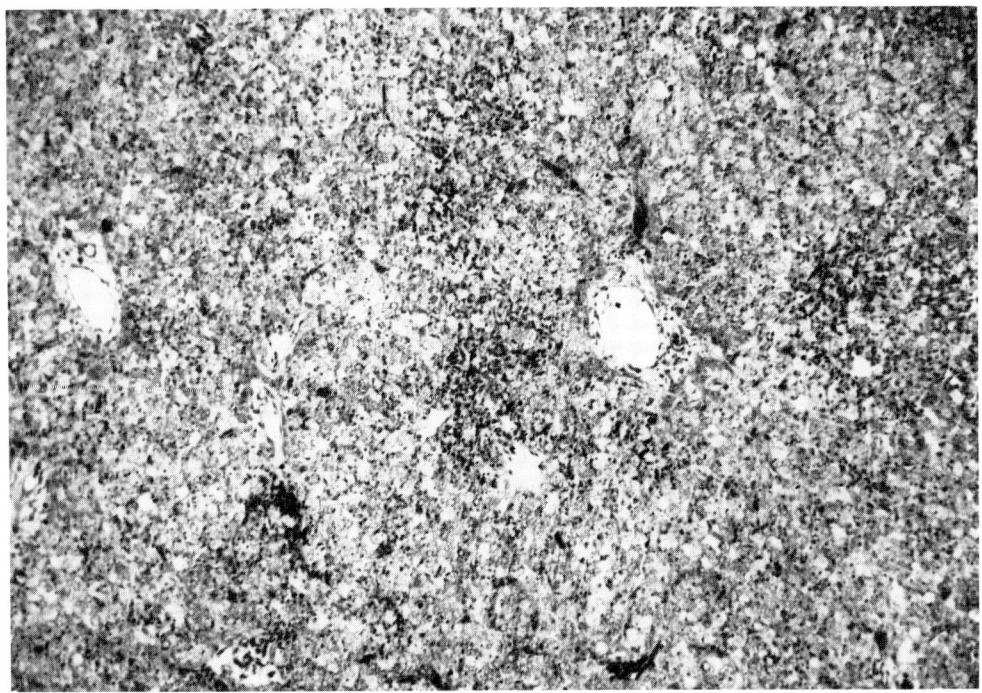

FIGURE 25. Fatty change, hemorrhages, and necrosis in duckling liver 24 h after administration of the substance isolated from the patient's liver (hematoxylin and eosin, ×131).

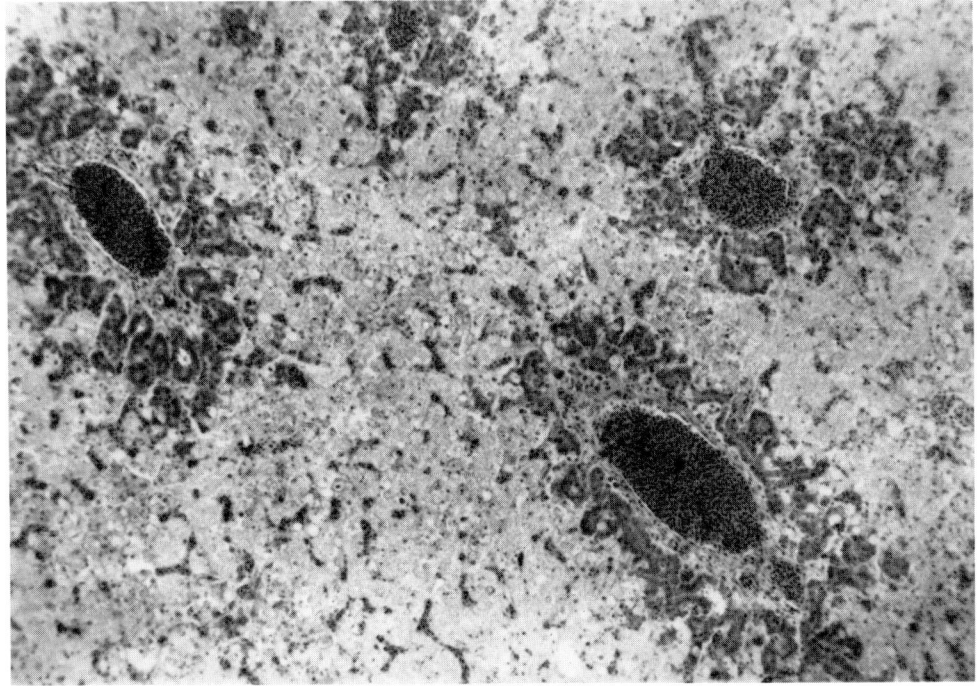

FIGURE 26. Severe bile duct proliferation of duck liver 10 d after administration of the substance isolated from patient's liver (hematoxlyin and eosin, ×88).

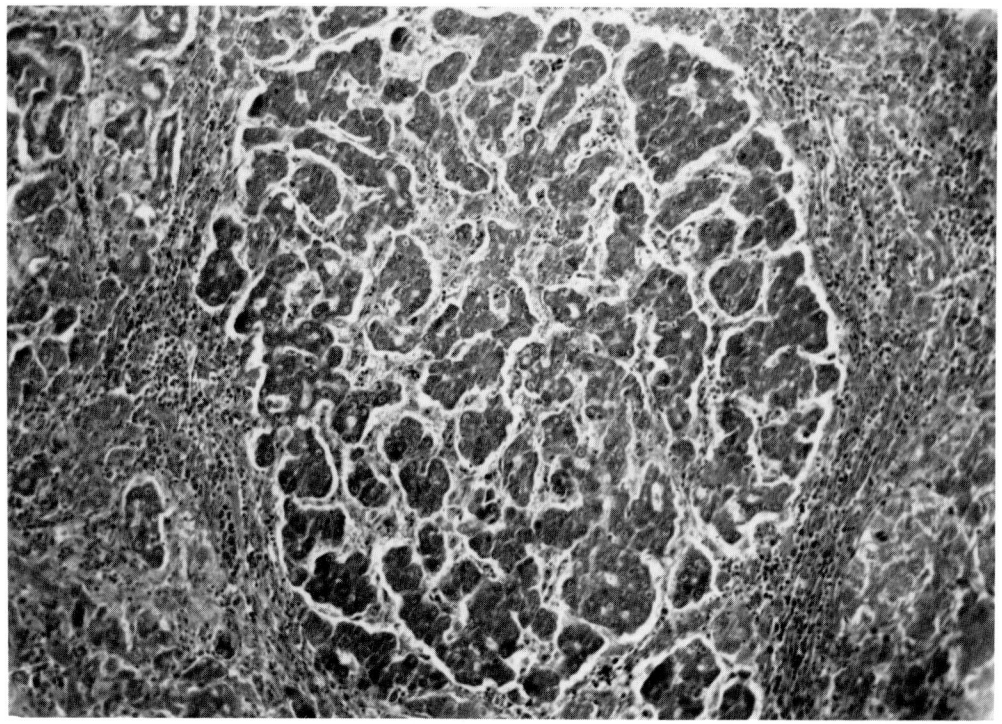

FIGURE 27. Cirrhosis with hyperplastic nodules of duck liver 4 weeks after administration of the substance isolated from patient's liver (hematoxylin and eosin, ×90).

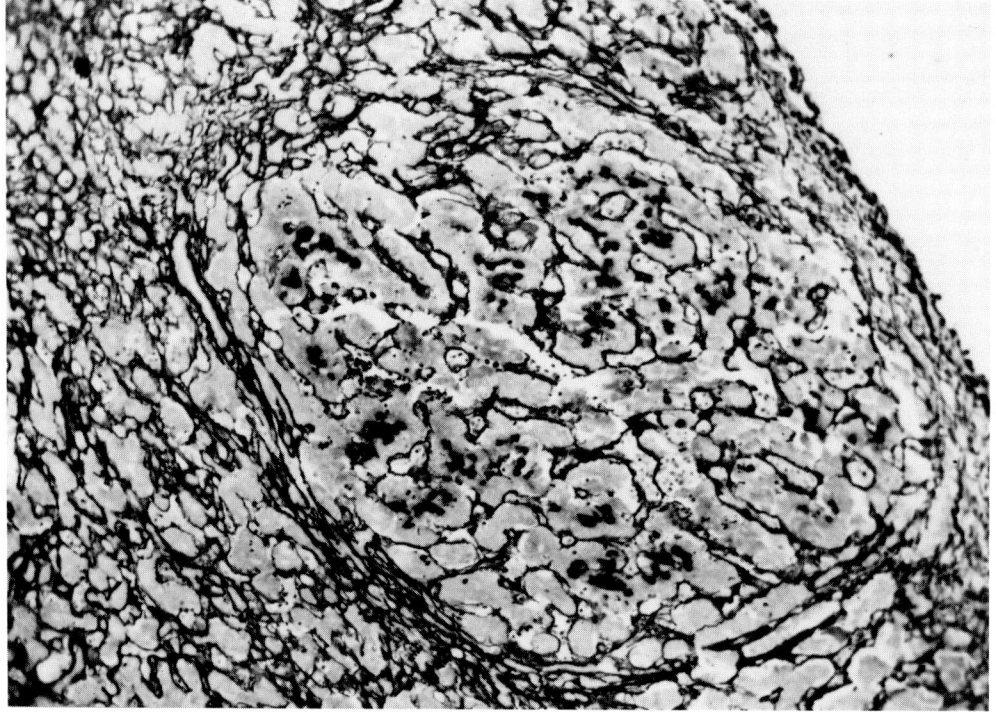

FIGURE 28. Hyperplastic nodules and proliferation of the reticulin fibers of the duck liver (Gomori, ×135).

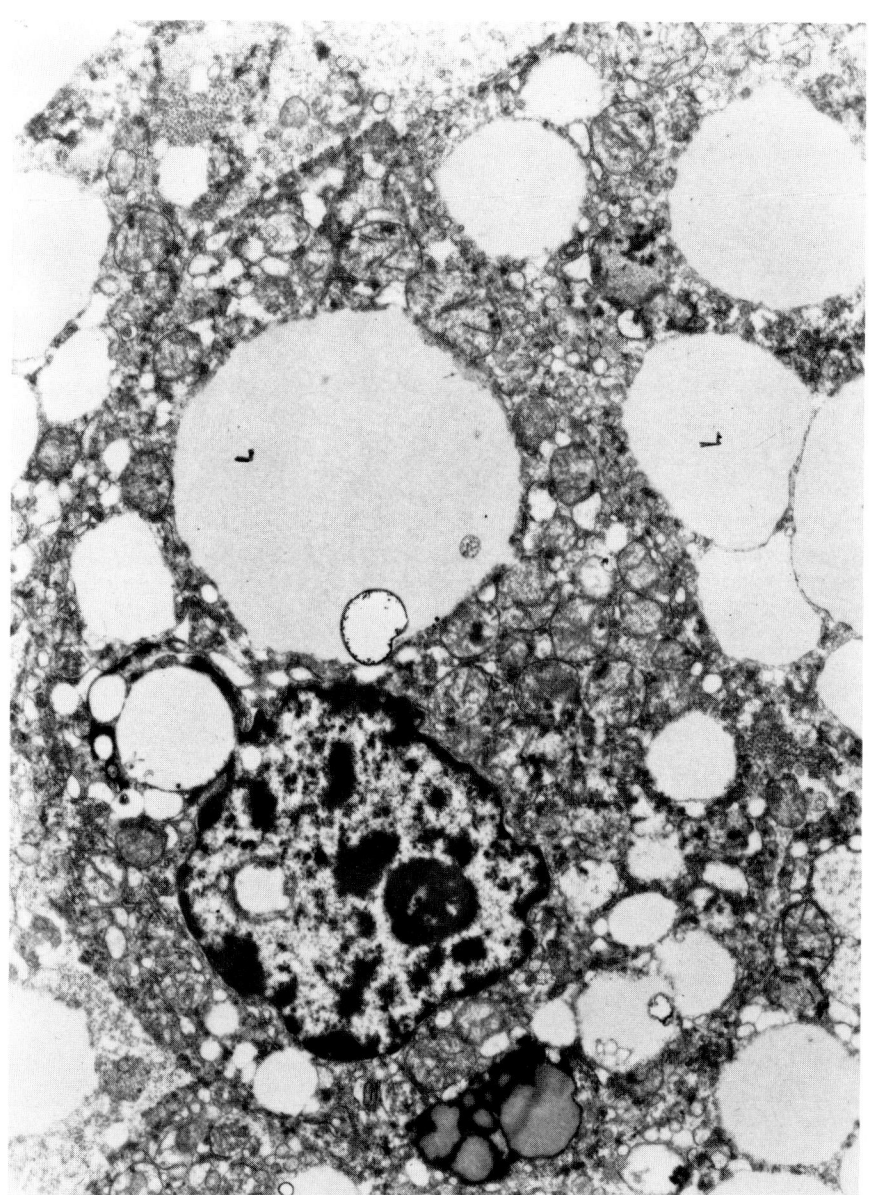

FIGURE 29. Electron micrograph of liver biopsy showing many lipid droplets (L) and swollen mitochondria in the cytoplasm (×6395).

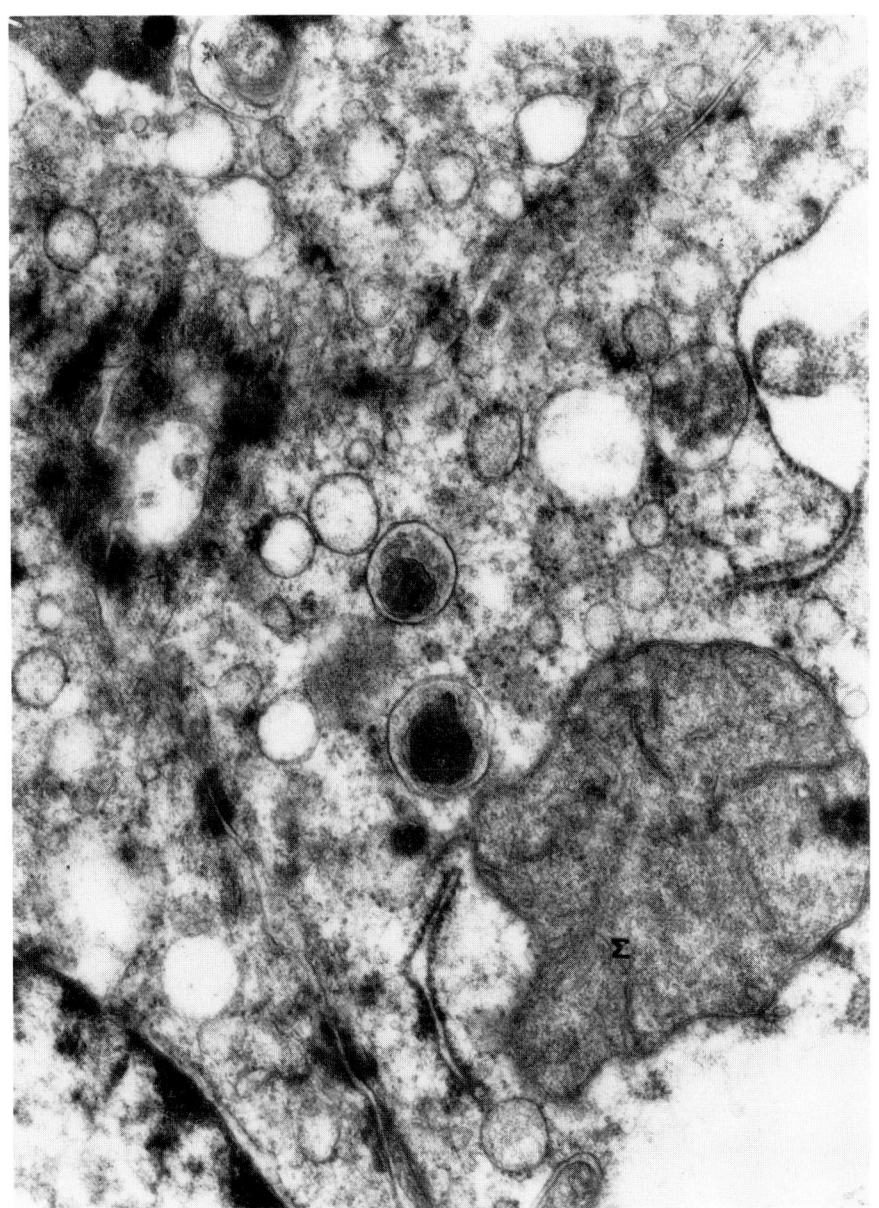

FIGURE 30. Swollen mitochondria (M) with disorganized matrix, reduction of the cristae, and loss of matrix dense bodies (liver biopsy, ×25,580).

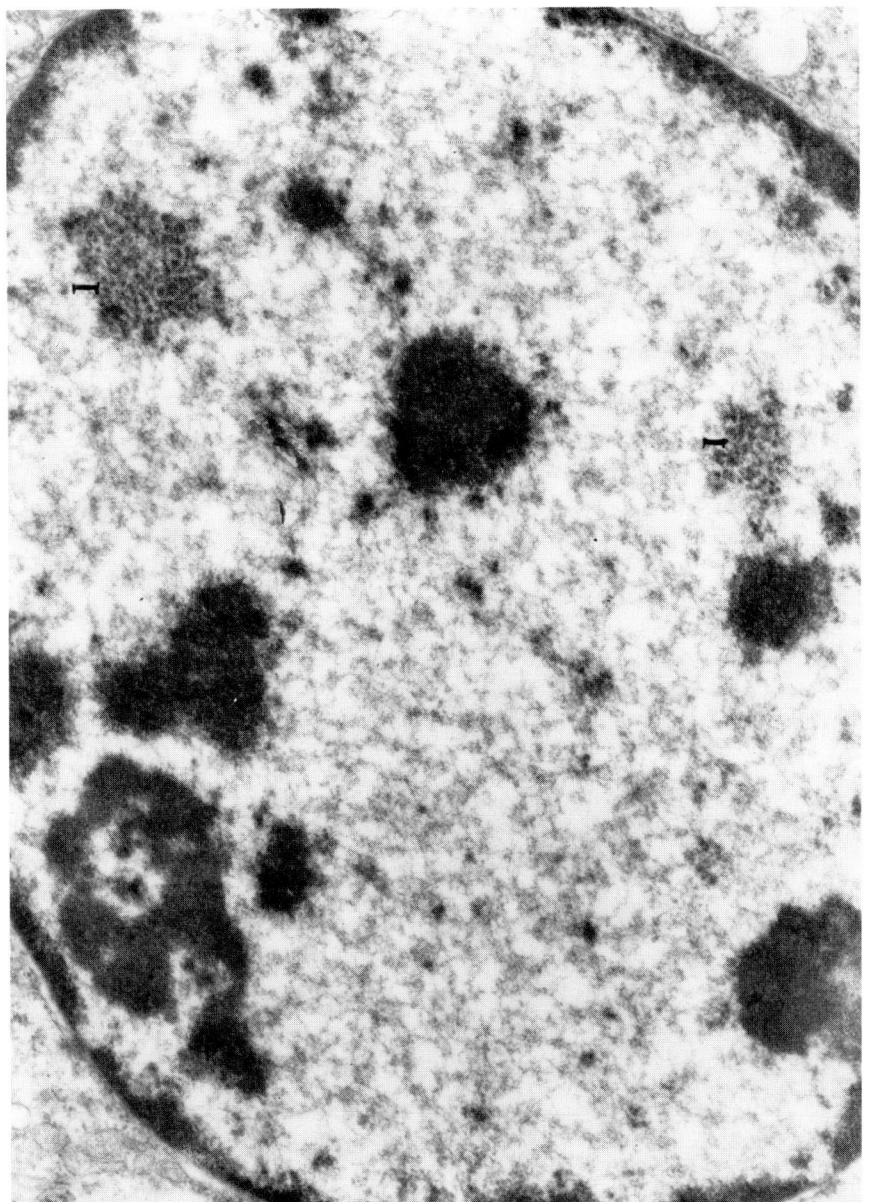

FIGURE 31. Clusters of granular material identical to interchromatine granules (I) are seen in the nucleus of the liver cell (liver biopsy, ×18,480).

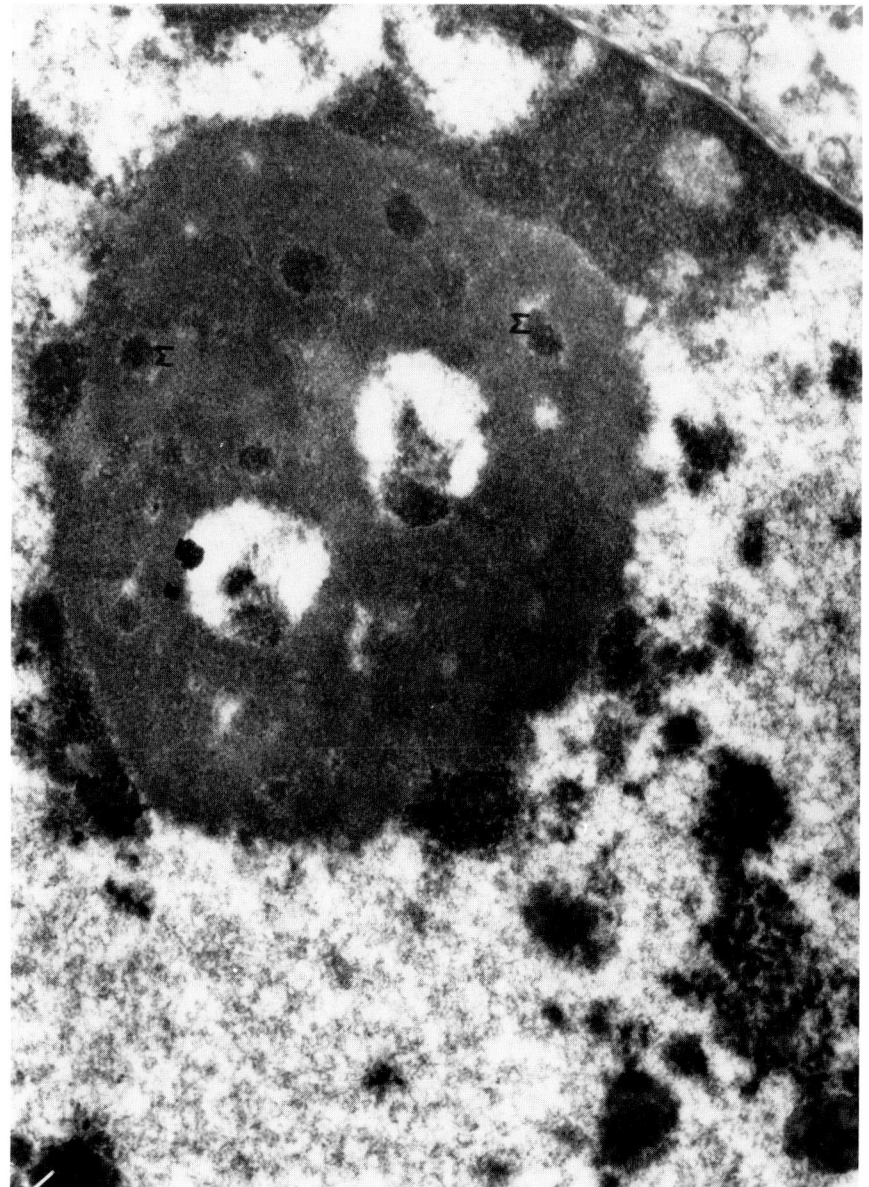

FIGURE 32. A giant nucleolus exhibiting a dense zone and a zone of moderate density containing vacuoles with opaque macrogranules (M). Islands of condensed chromatin along the inner surface of the nuclear membrane (liver biopsy, ×16,000).

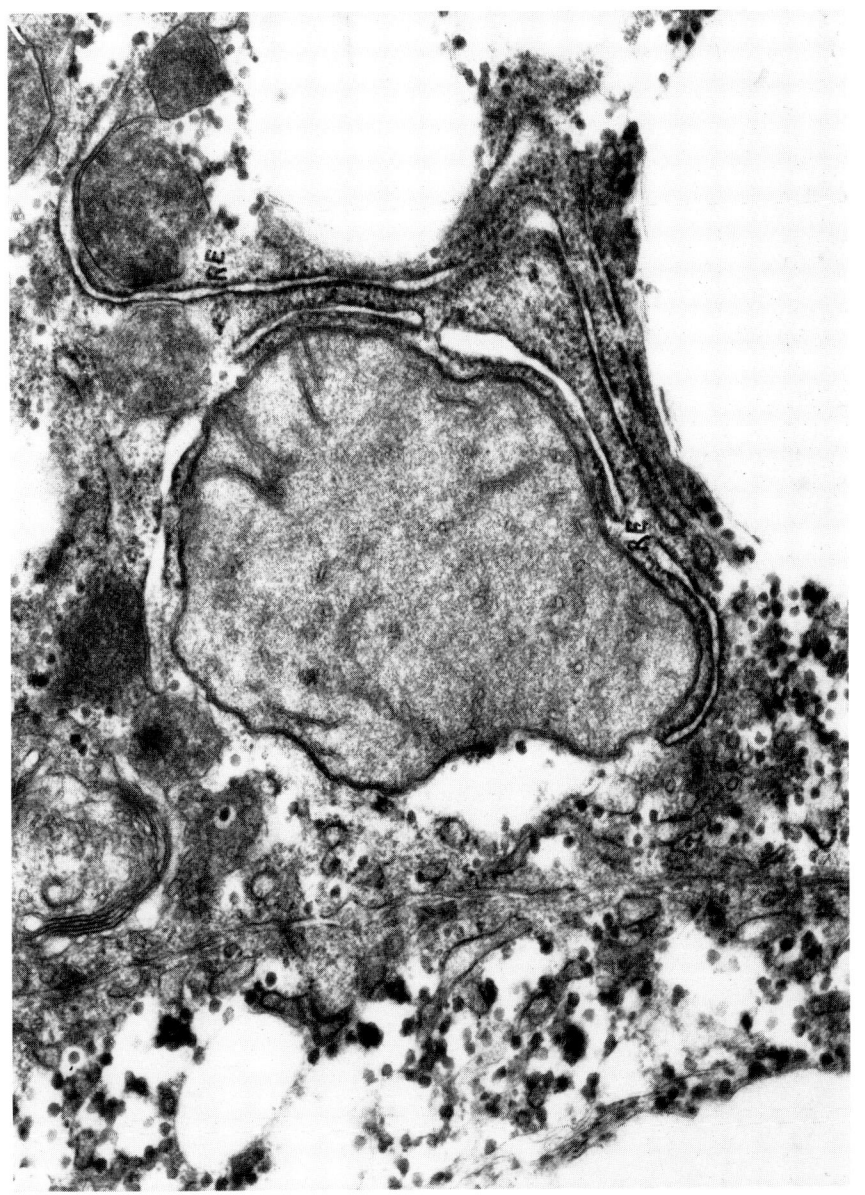

FIGURE 33. Swollen mitochondrion with loss of matrix dense bodies. Lamellae of the rough endoplasmic reticulum (RE) are seen in the vicinity of the mitochondrion (liver 2 h post-mortem, ×27,840).

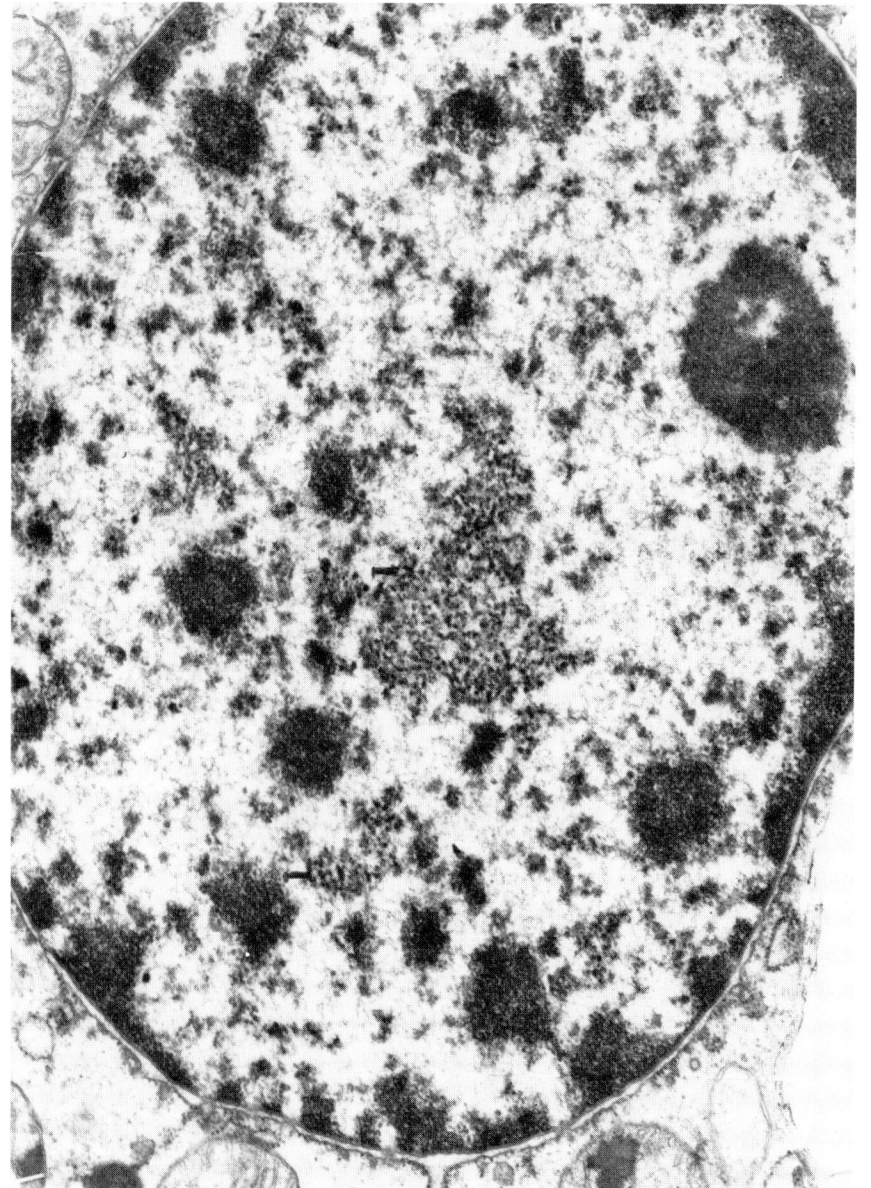

FIGURE 34. Nucleus of the liver cell with clusters of interchromatine granules (I) (liver 2 h post-mortem, ×18,270).

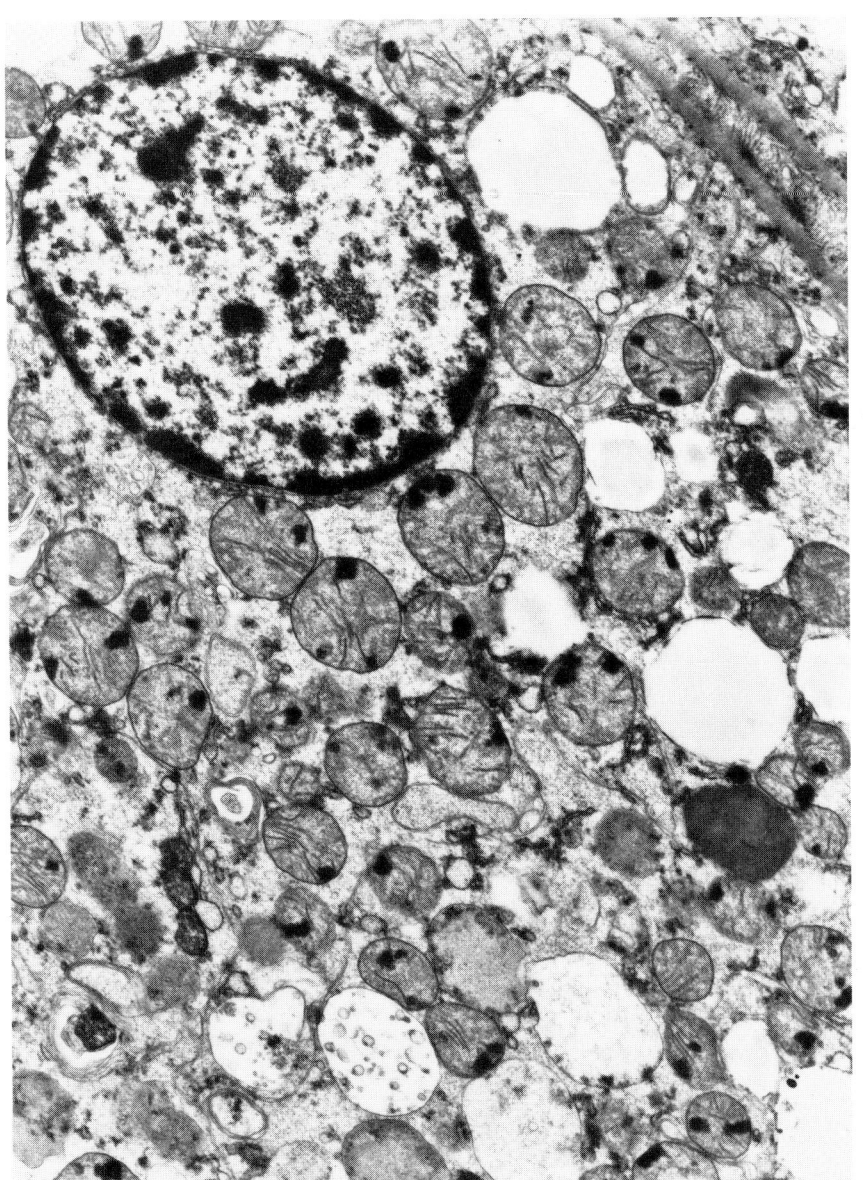

FIGURE 35. Tubular cell of the kidney with lipid vacuoles and swollen mitochondria. Matrix dense bodies are present (kidney 2 h post-mortem, ×6966).

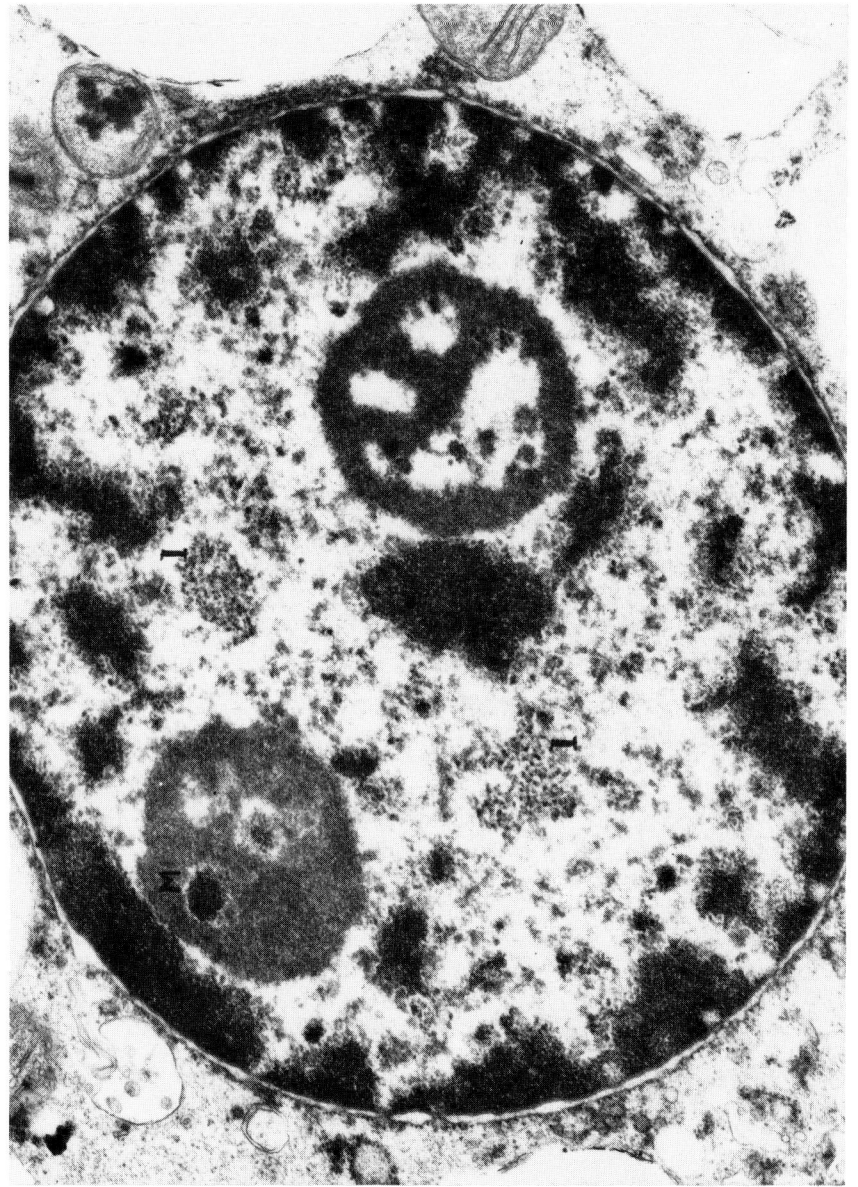

FIGURE 36. Nucleus of the tubular cell with two nucleoli containing vacuoles with macrogranules (M). Clusters of interchromatine granules (I) are seen in the nucleoplasm. Islands of condensed chromatin along the inner surface of the nuclear membrane (×18,060).

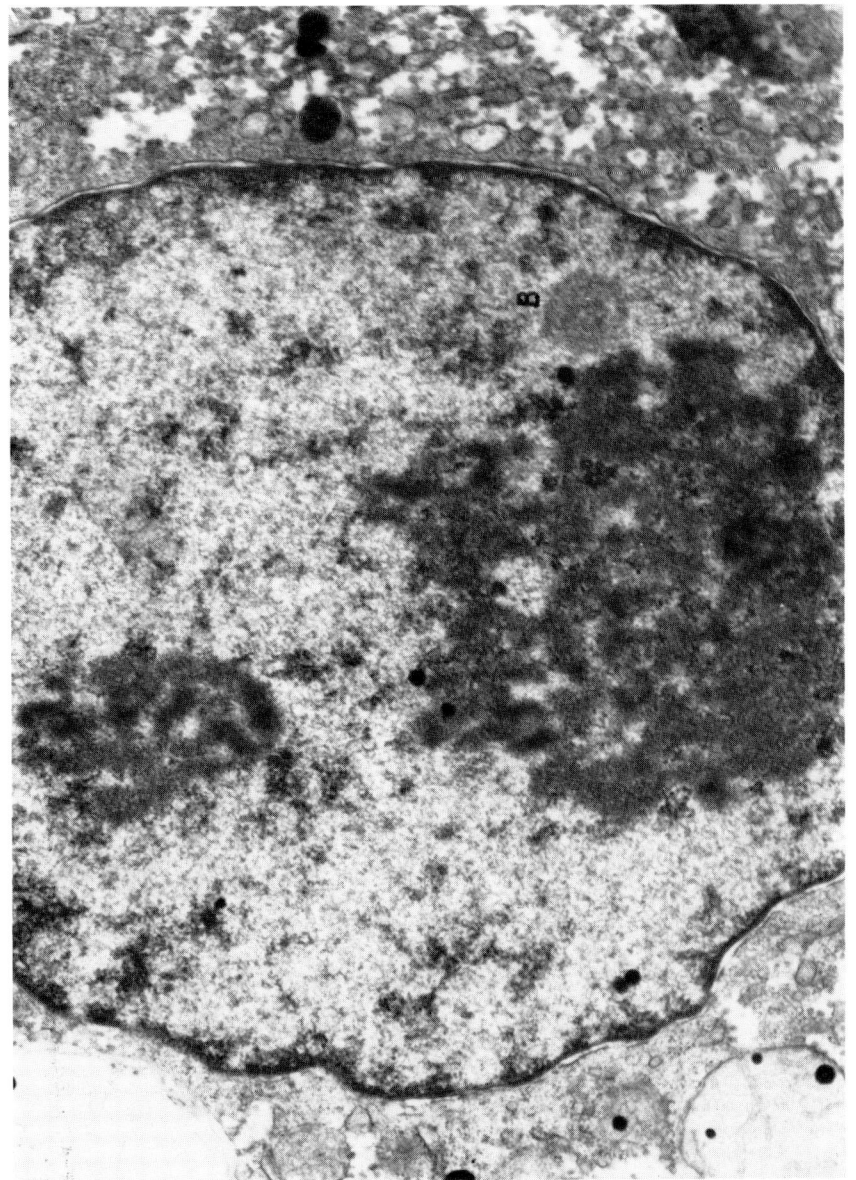

FIGURE 37. Nucleolar body (B) adjacent to the enlarged nucleolus of the tubular cell of the kidney (×19,780).

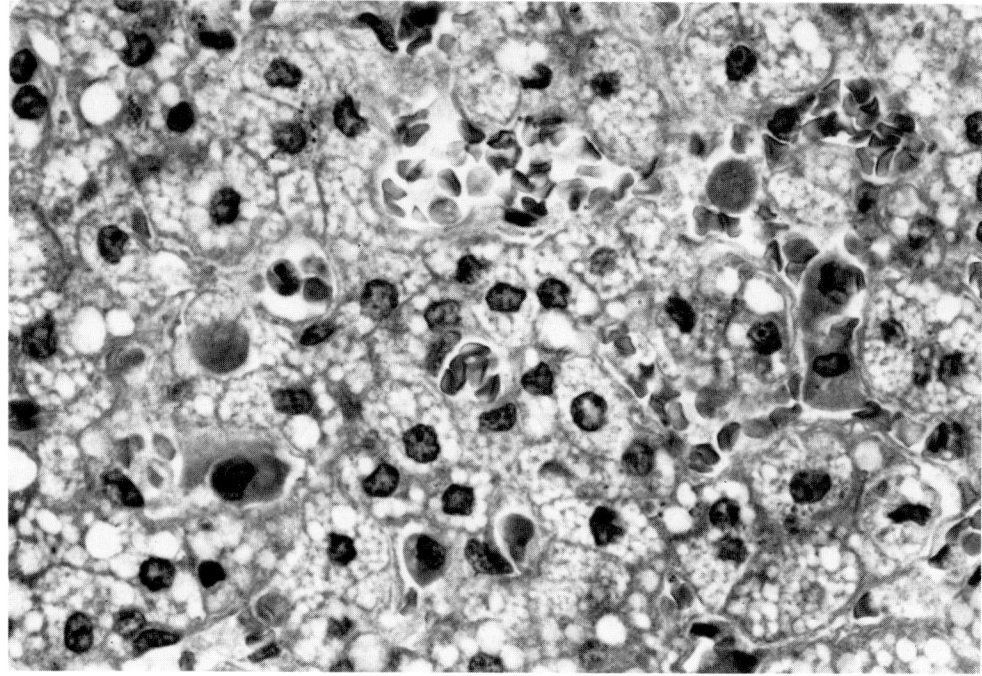

FIGURE 38. Diffuse fatty change in the liver biopsy taken on the third day of the illness (Table 22, case 2, hematoxylin and eosin, ×174).

The gross pathological findings were cerebral edema, enlarged yellow liver of firm consistence, and hyperplasia of the mesenteric lymph nodes. Virological and bacteriological investigations were negative. Serological investigations for the presence of HBsAg and hepatitis A infection also gave negative results.

The histological examination revealed cerebral edema with degenerated or necrotic cortical neurons and numerous swollen astrocytes throughout the brain. No inflammatory reaction was present. A fatty change was observed in the kidneys, myocardium, and striated muscles, while fat-laden histiocytes were present in the spleen and lymph nodes.

In three children, two girls and one boy (cases 1 to 3), the liver architecture was preserved, but portal tracts were distended and fibrous tissue infiltrated with monocellular infiltrates penetrating into the parenchyma. A marked bile-duct proliferation was seen. The parenchymal cells contained a large amount of fat dispersed in small vacuoles (Figure 39, 40).

In two boys and one girl (cases 4 to 6), the lobular pattern of the liver was disorganized and irregular regenerative nodules were found composed of basophilic hepatocytes with fine fat droplets on their venous poles (Figures 41 and 42). The remaining parenchymal cells were filled with fat vacuoles which contained a great amount of fine doubly refractive crystals.

In the liver samples of five patients (in one case the material was not available), AFB_1 at concentrations of 30 to 1126 µg/kg of the liver was chemically proved.

Samples of food consumed by a 14-month-old boy at the onset of the disease were investigated for the presence of aflatoxin. AFB_1 was found at the level of 310 µg/kg in one sample of powdered cocoa milk.

The clinical, laboratory, and morphological findings in the organs were consistent with the changes found in patients with an acute course of the disease, but were different in the livers. The morphological changes in the livers of these patients most probably represented a developmental sequence of the changes from steatosis to cirrhosis. This assumption was

TABLE 28
Reye's Syndrome Cases with Chronic Liver Damage

Case no.		Sex	Age (months)	Prodromal symptoms	Vomiting	Convulsions	Improvement (within days)	Transaminases (mkat/l)		Pareses	Death (within months)	Aflatoxin (µg/kg)	
								AST	ALT			Liver	Food
1.	I.K.	F	8 m	URI	+	+	5	3.40	3.28	+	2	90	—
2.	V.Č.[a]	M	8 m	URI	+	+	7	4.30	4.50	+	3	70	—
3.	L.D.	F	3 m	URI	+	+	3	—	—		2	30	—
4.	V.K.	F	12 m	URI	+	+	10	3.70	3.31	+	4	—	—
5.	M.J.	M	14 m	URI	+	+	7	3.10	3.30	+	3	1126	510
6.	J.K.	M	14 m	URI	+	+	6	—	—	+	3	102	—

Note: URI = upper respiratory infection.

[a] Liver biopsy performed on the third day of the illness.

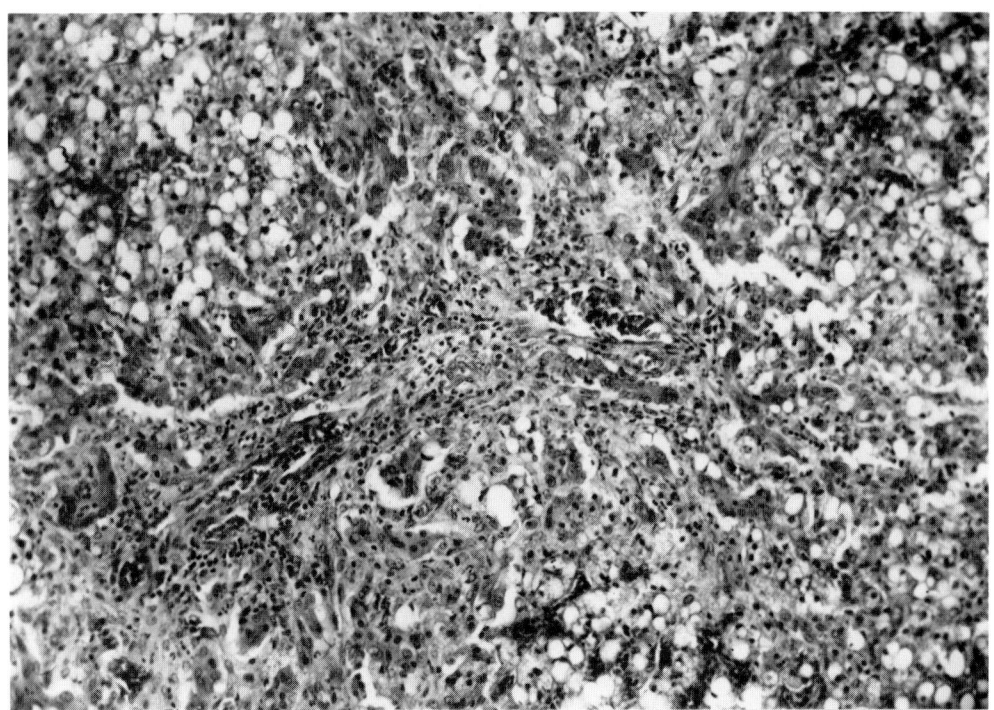

FIGURE 39. Periportal fibrosis with bile duct proliferation. The fibrous tissue is infiltrated with monocellular infiltrates. The parenchymal cells contain numerous fat vacuoles (Table 22, case 2, hematoxylin and eosin, ×132).

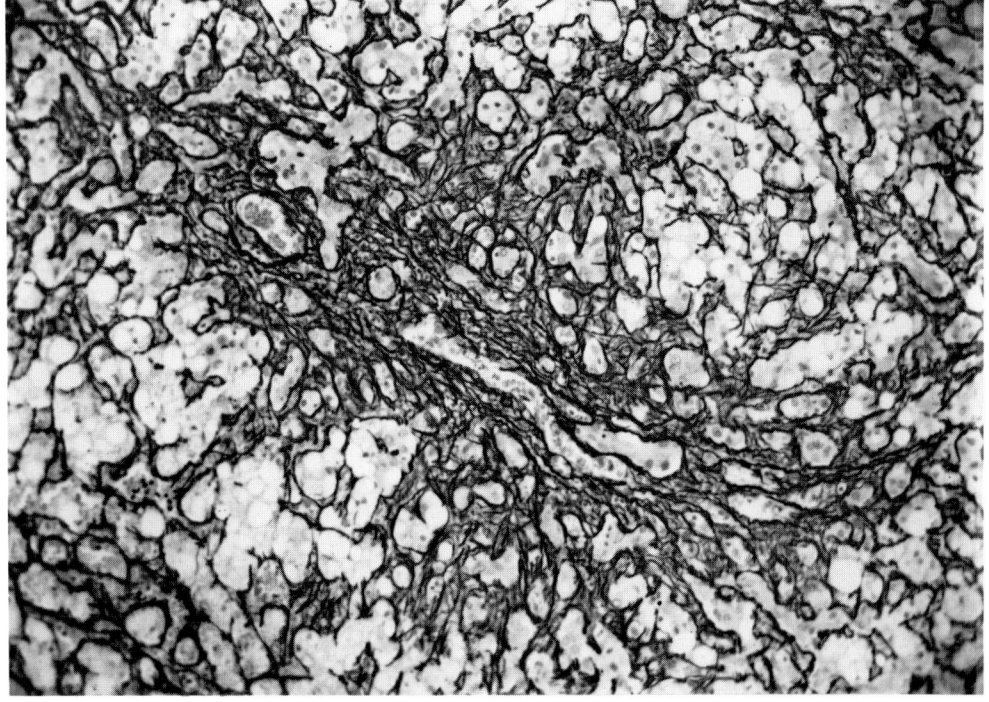

FIGURE 40. Severe proliferation of the reticulin fibers in the portal tract (Table 22, case 2, Gomori, ×132).

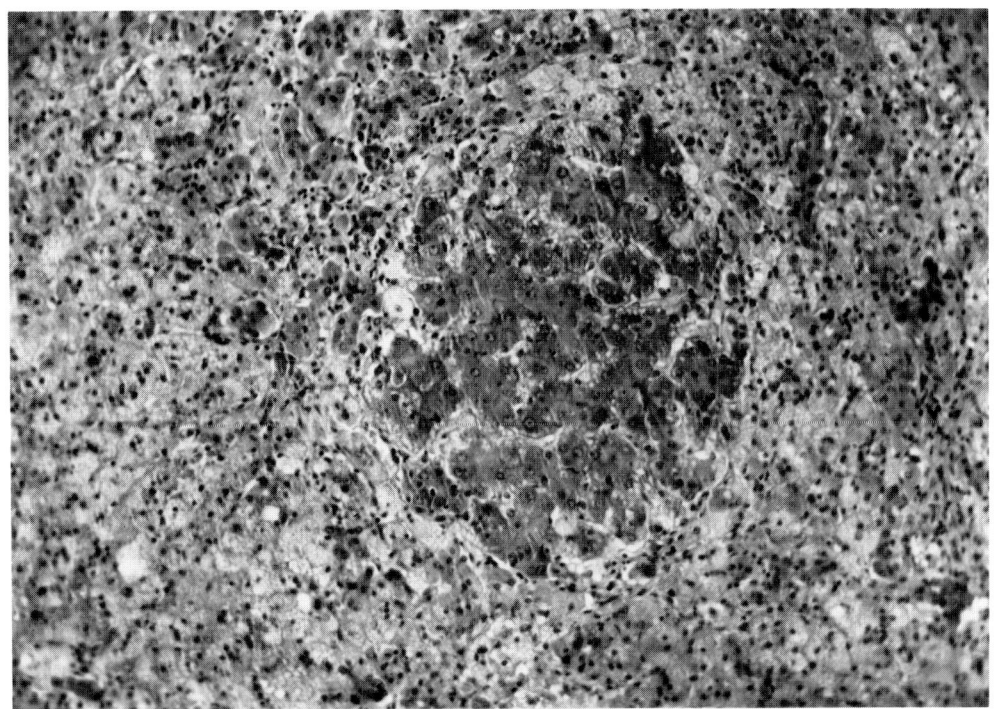

FIGURE 41. Hyperplastic nodule and fine droplet steatosis of the liver cells (Table 22, case 5, hematoxylin and eosin, ×88).

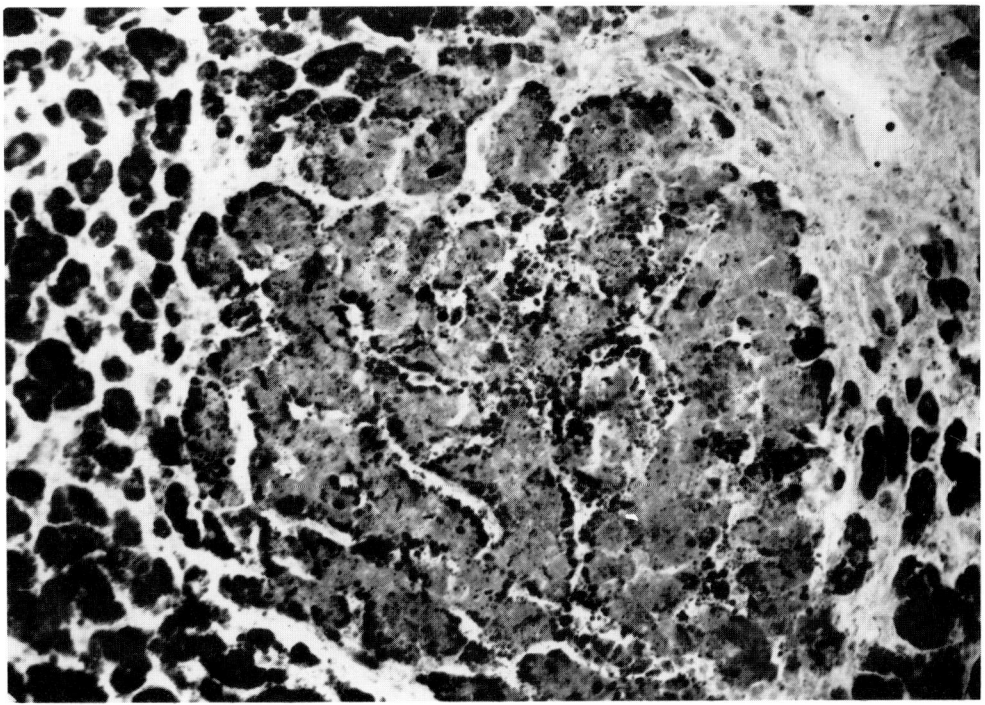

FIGURE 42. Hyperplastic nodule with fine fat droplets on the venous poles of the cells. The remaining parenchymal liver cell show diffuse fatty change (Table 22, case 5, Sudan, ×131).

supported by finding a diffuse steatosis at the liver biopsy of an 8-month-old boy, in whom liver fibrosis with bile-duct proliferation was found 2 months later.

The presence of AFB_1 in the liver samples of these patients was of special interest and could indicate a possible link between chronic liver damage and aflatoxin. A similar observation has been reported by Amla et al.[12] in children aged 1.5 to 15 years in India, who were known to have consumed aflatoxin-contaminated food during specific time periods. Liver biopsies revealed a fatty degeneration during the first 2 months, later followed by fibrosis and cirrhosis.

Becroft and Webster[13] reported an 8-month-old girl who developed clinical and laboratory symptoms of RS at the age of 5 months and died 3 months later. The histological examination revealed a severe bile-duct proliferation with a diffuse fatty change in her liver, and in the liver extract, AFB_1 in an amount of 50 µg/kg wet weight was found.

In addition, the liver changes in our patients closely resembled those observed in the biologically tested, aflatoxin-fed ducklings, as reported earlier.

The discovery of aflatoxin in a chronically damaged liver was surprising and could indicate either its long-term persistence in the liver tissue or a recent exposure of patients to aflatoxin. The study of Dalezios and associates[14] has shown that 80 to 85% of the orally administered dose of labeled AFB_1 to rhesus monkeys was excreted in the urine and feces within 7 d; however, their livers still retained about 1% of the dose 5 weeks after the administration. The rat liver was also found to retain appreciable amounts of AFB_1 for a long time.[14]

A recent exposure of children to aflatoxin seems less likely since the children were nourished almost exclusively parenterally during the last weeks in the hospital. Thus the persistence of the toxin in the liver tissue seems more likely, although its high concentration is difficult to explain. It is known that aflatoxin is primarily metabolized by liver microsomal mixed-function oxidases, resulting in a variety of detoxification and activation products. The hypothesis that a deficient liver enzyme system could be responsible for the retention of unmetabolized AFB_1, or that the pathological changes connected with RS could have decreased the clearance of aflatoxin from the tissue, has been suggested.

V. REYE'S SYNDROME IN THE NEWBORNS

Despite medical awareness and a great number of case reports, RS in the neonatal period is quite unusual. In world literature, only two cases were published until 1976. The first case reported by Papageourgiou and collaborators[16] was a 4-d-old girl from an uncomplicated pregnancy. The infant was well until 40 h of age when tachypnea and spastic episodes were noted. Arterial blood gases and acid-base studies revealed a metabolic acidosis with a pH of 7.26. Blood sugar was 14 mg/100 ml; CSF contained 800 mg/100 ml of protein and no sugar. The post-mortem examination showed a fatty change of the liver, kidney, and myocardium. Sections of the brain showed edema and swelling of astrocytes. Bacteriological examination of the blood, lungs, and spleen was negative. The diagnosis of RS was suggested in view of the morphological changes. The authors assumed that some intrauterine event, either genetic or due to an unidentified toxin or infectious process, was responsible for the immediate post-natal course.

The second case reported in 1976 by Harris et al.[17] was a 3400 g female infant delivered after 40 weeks of uncomplicated gestation. The infant did well until 24 h after birth when respiratory distress was noted. At 62 h of age, tonic seizures and temperature instability were documented. The electroencephalogram was abnormal with a run of high voltage on the left side. Transaminases and ammonia were elevated, and the prothrombin time was prolonged. Metabolic acidosis with partial respiratory compensation was noted. Exchange transfusions were performed and subsequently the baby improved. Her physical and development examination at 7 weeks of age was normal. A follow-up study of this patient at 5 months of age suggested minimal residual hepatic damage with a slightly enlarged liver and an elevated SGOT level, but normal neurologic status. The authors postulated that the clinical feature of the illness, though slightly different in that the respiratory distress was presenting signs with no mild preceding illness or vomiting, fulfilled the established clinical, nonhistological criteria for RS.

During the follow-up study of RS cases at the Department of Pathology in Hradec Králové, six newborns have been observed who fulfilled the morphological criteria for this disease. With regard to the rarity of this illness in the neonatal period, the case history is presented in more detail.

Case 1 — K.J., a baby boy from the third uncomplicated pregnancy weighing 1550 g, was born in the 32nd week of gestation to a 31-year-old gravida. Two previously born brothers (6 and 7 years) were healthy. The mother worked on a cooperative farm. The infant did well after delivery, but at 12 h after birth he developed apneic spells and marked hypotonia. Arterial blood gases and acid-base studies revealed a severe metabolic acidosis (base excess –14 with a pH of 7.12). The infant died within 32 h with the clinical diagnosis of "Respiratory Distress Syndrome" (RDS). The autopsy showed partial atelectases of the lung, a fatty liver, and brain edema.

Case 2 — V.M., a girl born in the 30th week of the first uneventful pregnancy, weighing 2150 g. The child's mother was a clerk, and the family lived in a small city. Shortly after delivery, respiratory distress and extreme acidosis with base excess –12 and a pH of 7.19 were noted. Intensive bicarbonate alcalic therapy was introduced without any effect. The child died 67 h after birth with the clinical diagnosis of immaturity and RDS. The autopsy findings showed a subdural hematoma, atelectases of the lungs, a pale liver, and a slight flattening of the brain gyri.

Case 3 — V., a premature male weighing 1330 g was born in the 29th week of the first pregnancy. The child's mother worked in a cooperative farm as a feeder, and the family lived in the country. The infant was well and normally active until he was 20 h of age when tachypnea, acidosis, and respiratory alteration and hypotonia were noted. Intensive alcalic therapy was without any effect. Shortly before death, the infant vomited blood and died at 75 h of age with the clinical diagnosis of immaturity and RDS. The autopsy findings showed hematocephalus and a yellow liver.

Case 4 — V.M., a male born in the 30th week of the first pregnancy weighing 1330 g. The mother was a laboratory technician on a cooperative farm. The family lived in the country. Four subsequent siblings (three boys and one girl) were healthy. Respiratory distress and a severe acidosis (base excess –13, pH 7.1), tachypnea, and hypotonia were noted 10 h after birth. The blood-sugar level was 20 mg/100 ml. The infant died within 70 h with the clinical diagnosis of RDS. The autopsy findings showed a small subdural hematoma, atelectases of the lungs, and a yellow liver.

Case 5 — V.K., a female born in the 40th week of gestation from the third normal pregnancy, weighing 3200 g. Two boys from the first and second pregnancy were healthy. The mother worked on a cooperative farm as a feeder, and the family lived in the country. The child was well and normally active after birth. On the second day, 2 h after breast feeding, the infant became pallid and had frequent apneic spells, and died 45 h after delivery with the clinical diagnosis of sudden infant death syndrome (SIDS). The autopsy revealed brain edema, a yellow liver, and a yellow tinge to the cortex of the kidney and heart.

Case 6 — V.O., a full-term male (3400 g) from the fifth normal pregnancy (a brother of the female born in case 5). The infant did apparently well until 48 h after birth, when pallor, temperature instability, and a heart disorder were noted. The infant died at 72 h of age with the clinical diagnosis of SIDS and a suspected heart disease. The autopsy revealed a yellow liver, brain edema, a yellow tinge to the cortex of the kidneys, heart, and striated muscles. The boy from the mother's fourth pregnancy was healthy.

The familial and clinical data are presented in Tables 29 and 30. A histological study of all the six infants showed pericellular and perivascular edema of the brain cortex (Figure 43), a diffuse fatty infiltration of the liver (Figure 44), and a fatty change of the tubular cells of the kidneys (Figure 45). The myocardium revealed large areas of fine lipid droplets (Figure 46). Fibers of the striated muscles showed granular decomposition resembling Zenker's degenera-

TABLE 29
Reye's Syndrome in the Newborns Familial Data

Case no.	Sex	Age (hours)	Year of death	Urban area	Rural area	Employment of mother	Pregnancy Pregnancy	Brothers and sisters
1. K.J.	M	32	1972		+	Co-op farm worker	III	2
2. V.M.	F	67	1972	+		Clerk	I	—
3. V.J.	M	75	1972		+	Co-op farm worker	I	—
4. V.M.	M	70	1972		+	Co-op farm technician	I	4
5. V.K.[a]	F	45	1972		+	Co-op farm worker	III	3
6. V.O.[a]	M	72	1976		+	Co-op farm worker	V	3

[a] Case 5 and 6: brother and sister.

TABLE 30
Reye's Syndrome in the Newborns Clinical Data

Case no.	Sex	Birth weight and length	Clinical symptoms	Clinical diagnosis	Death after birth (hours)
1. K.J.	M	1550/40	Hypotonia Acidosis pH 7.12	Immaturity	32
2. V.M.	F	2150/42	Tachypnoe Acidosis pH 7.19	Immaturity RDS	67
3. V.J.	M	1330/37	Hypotonia Acidosis pH 6.9	Immaturity RDS	75
4. V.M.	M	1330/41	Hypotonia Acidosis pH 7.1 Blood sugar 20 mg/100 ml	Immaturity RDS	70
5. V.K.	F	3200/50		Sudden death	45
6. V.O.	M	3400/48	EKG-changes	Sudden death	72

tion and a fatty change of altered muscle fibers (Figures 47 and 48). Focal hemorrhages and numerous hyaline membranes were present in the lungs (Figure 49). The findings differed only in the intensity of fatty degeneration in various organs. The differences are shown in Table 31.

Samples of the psoas muscle and the diaphragm were processed for electron microscopy. Two kinds of changes were present. The first one, which corresponded with the granular disruption of muscle fibers in light microscopy, was a loss of cross-banding of the sarcomeres. Empty, membrane-free vacuoles (probably dissolved lipids) were present between the myofibrils. The second change observed was a homogenization of muscle fibers due to disintegration of A- and I-bands (Figures 50, 51).

Virological investigations of the heart for Coxsackie virus in cases 5 and 6 gave negative results.

Liver samples of three cases (cases 3, 4, and 6) were investigated for the presence of aflatoxin by chromatography and spectrophotometry. The liver sample extracts in cases 3 and 6 showed spots with blue fluorescence like that of commercial AFB_1 in 365 nm UV light, the same color change when treated with 40% H_2SO_4 and R_F values which are shown in Figure 52. In case 3, a quantitative analysis showed a concentration of 1015 µg/kg AFB_1. In addition to AFB_1, aflatoxin M_1 at a level of 9 µg/kg and two undefined metabolites X_1 and X_2 at a concentration of 450 µg/kg were detected. The liver sample of case 6 contained 80 µg/kg of AFB_1. The liver sample of case 4 was negative.

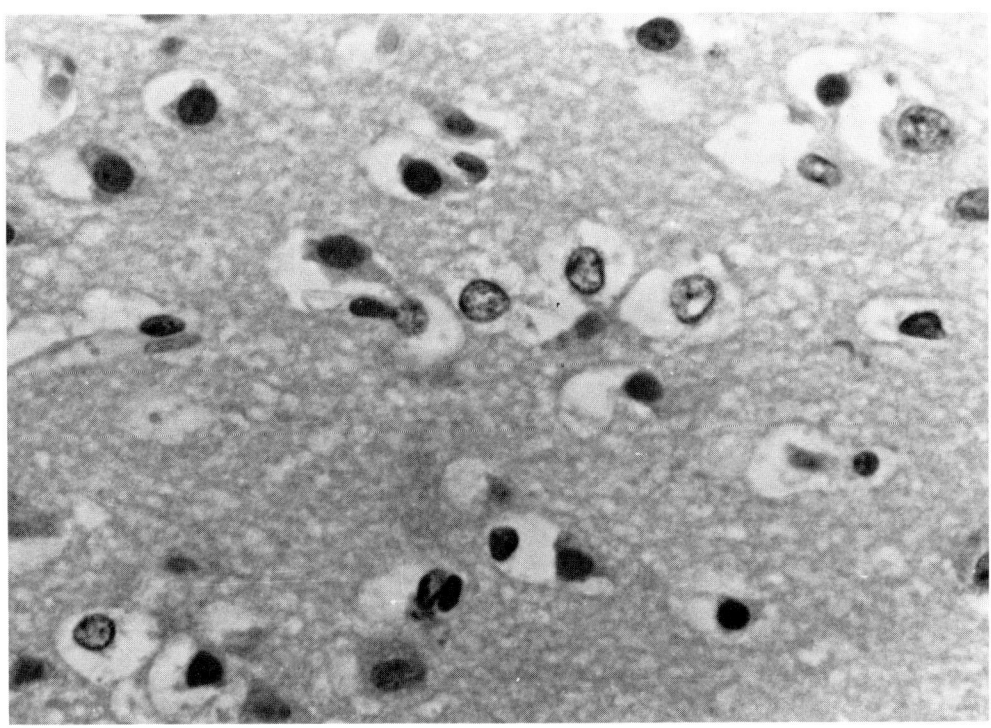

FIGURE 43. Brain edema and swollen astrocytes (Table 23, case 6, hematoxylin and eosin, ×135).

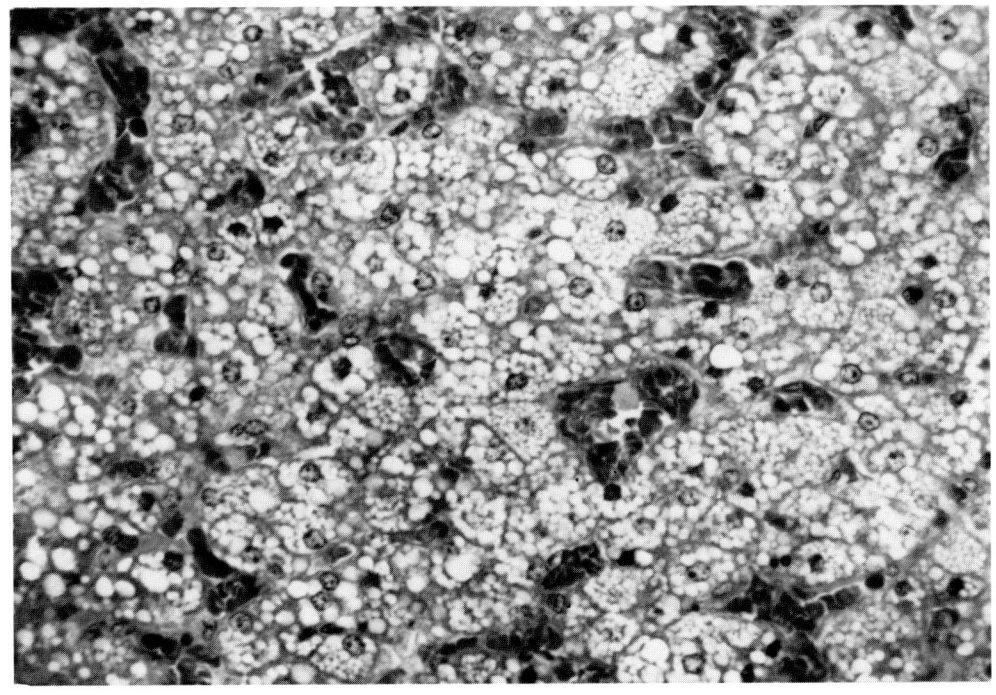

FIGURE 44. Diffuse fatty change of the liver (Table 23, case 3, hematoxylin and eosin, ×131).

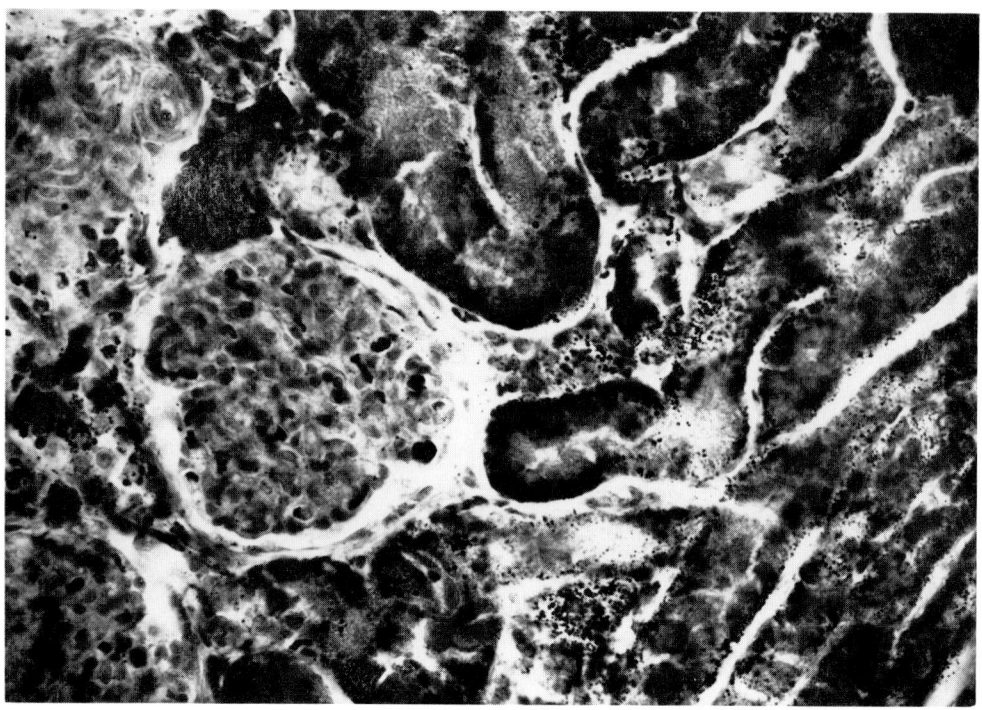

FIGURE 45. Fatty degeneration of the proximal tubules of the kidney (Table 23, case 6, Sudan, ×132).

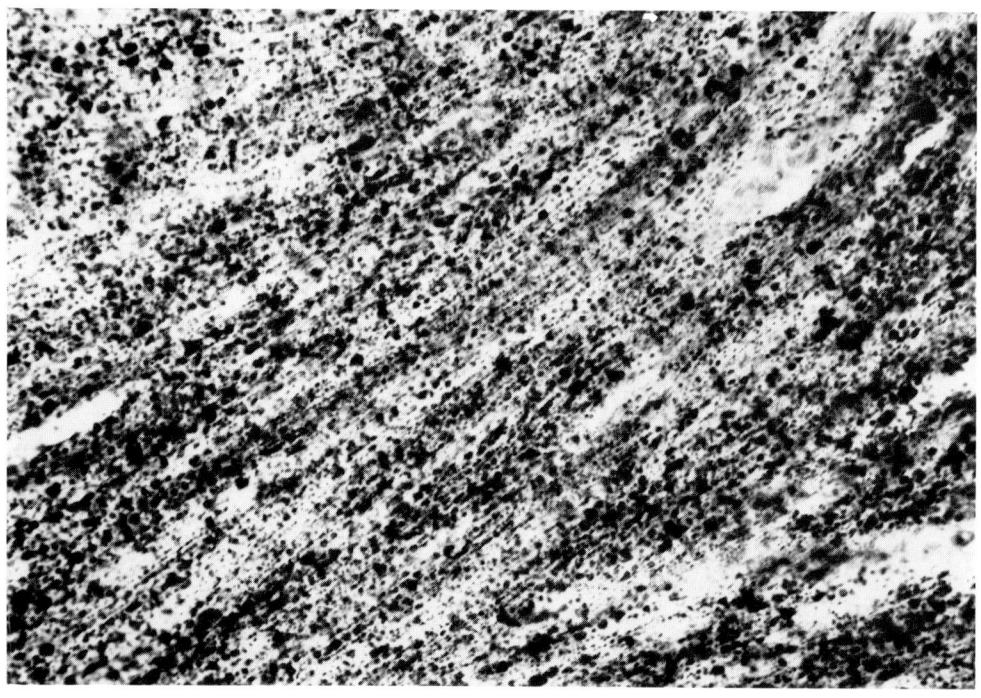

FIGURE 46. Fine droplet steatosis of the myocardium (Table 23, case 5, Sudan, ×131).

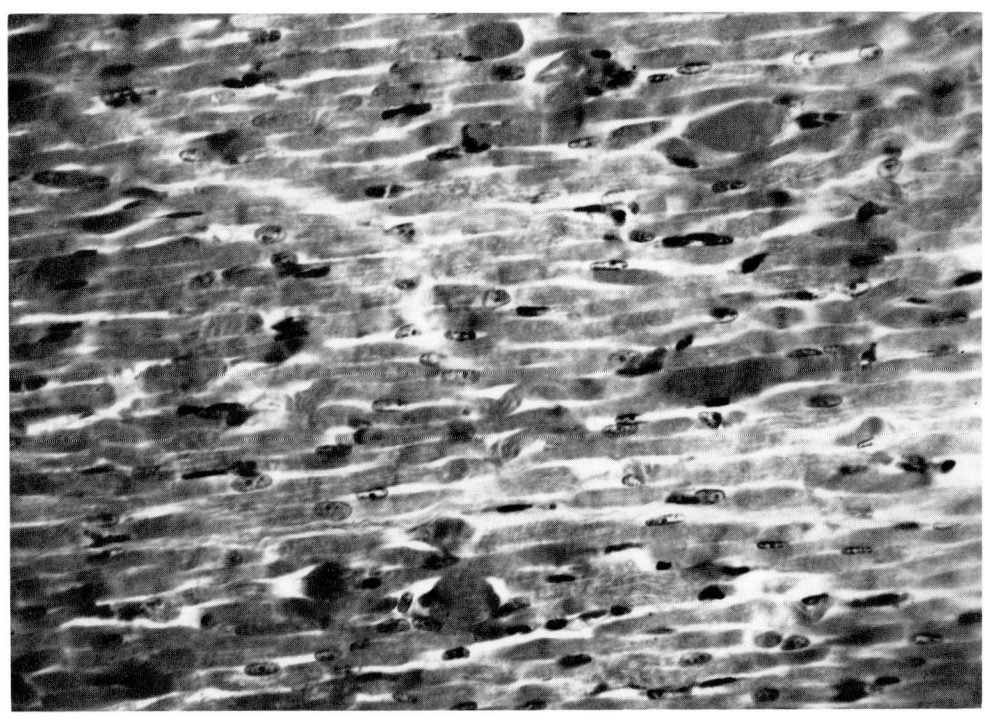

FIGURE 47. Granular degeneration of muscle fibers of the diaphragm (Table 23, case 6, hematoxylin and eosin, ×88).

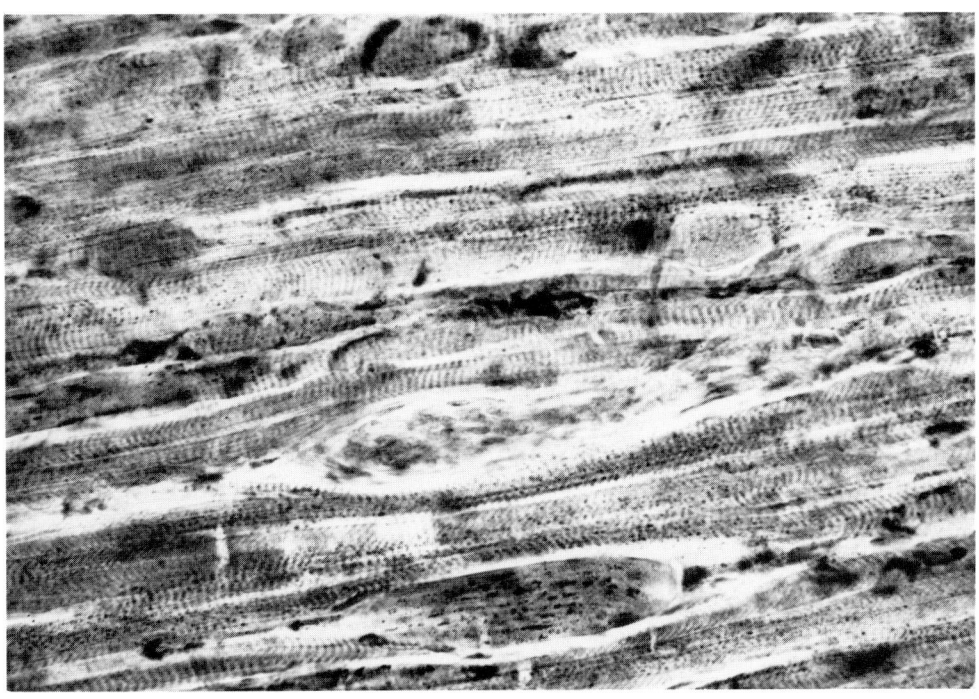

FIGURE 48. Fatty degeneration of altered muscle fibers (Table 23, case 6, Sudan, ×131).

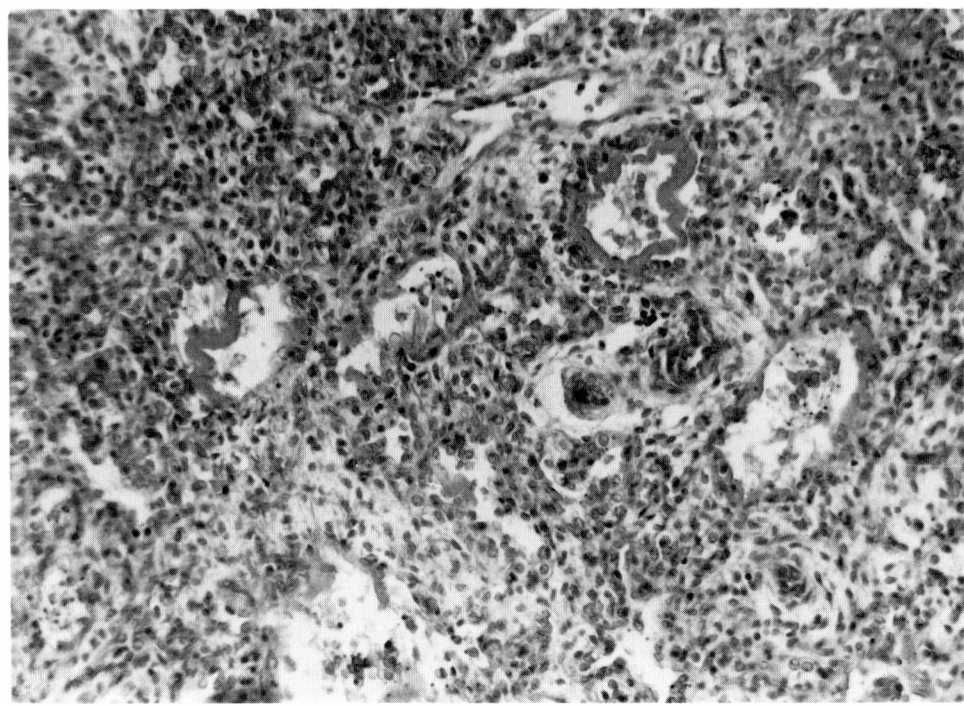

FIGURE 49. Hyaline membranes in the lung (Table 23, case 6, hematoxylin and eosin, ×131).

TABLE 31
Reye's Syndrome in the Newborns
Degree of Fatty Change

Case no.		Sex	Age (hours)	Liver	Kidney	Heart	Striated muscle
1.	K.J.	M	32	+++	+++	++	++
2.	V.M.	F	67	+++	++	++	+
3.	V.J.	M	75	+++	+++	+++	++
4.	V.M.	M	70	+++	+++	+++	++
5.	V.K.	F	45	+++	+++	+++	+++
6.	V.O.	M	72	+++	+++	+++	+++

The diagnosis of RS is usually based on the clinical course, biochemical alterations, and morphological changes characterized by fatty degeneration of viscera. Unlike the clinical picture in older children with RS, in these cases there was no prodromal respiratory infection, followed by the onset of vomiting and neurological symptoms. The clinical course in the first four premature newborns did not differ in its symptoms from other common cases of prematurity, in which respiratory distress is the initial abnormality noted. In these newborns, extreme acidemia without any improvement after alcalic therapy and a marked hypotonia were the remarkable signs.

The death of the siblings (cases 5 and 6) was sudden and unexpected. The diagnosis of RS in all the six infants was suspected based on the histological investigation. The morphological changes found in the organs — cerebral edema and fatty degeneration of the liver, kidneys, myocardium, and striated muscles — fulfilled the criteria for RS. Degenerative changes of the striated muscles were not an exceptional finding, but a common change observed in older

FIGURE 50. Electron micrograph of psoas muscle; loss of cross-banding. Empty membrane-free vacuoles between myofibrils (Table 23, case 6, ×28,980).

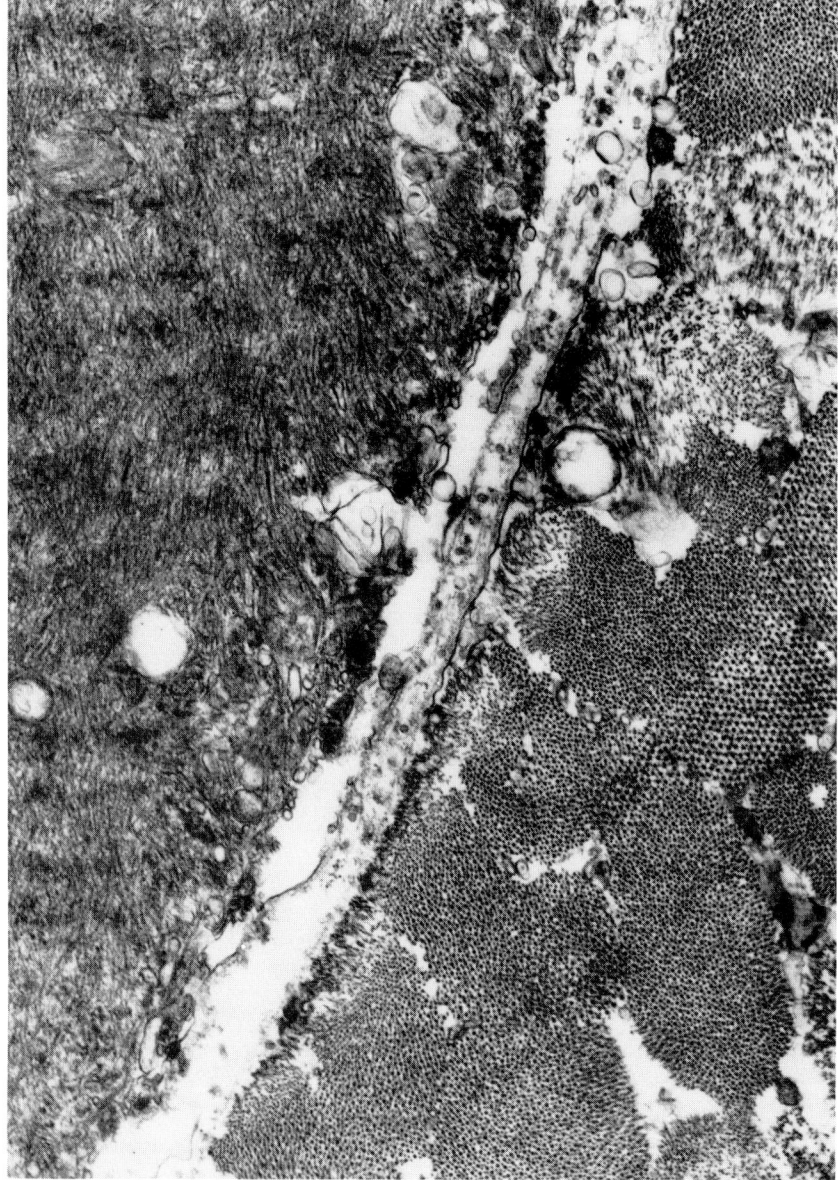

FIGURE 51. Electron micrograph of psoas muscle; homogenization of muscle fibers. Border between normal and altered muscle fibers (Table 23, case 3, ×28,980).

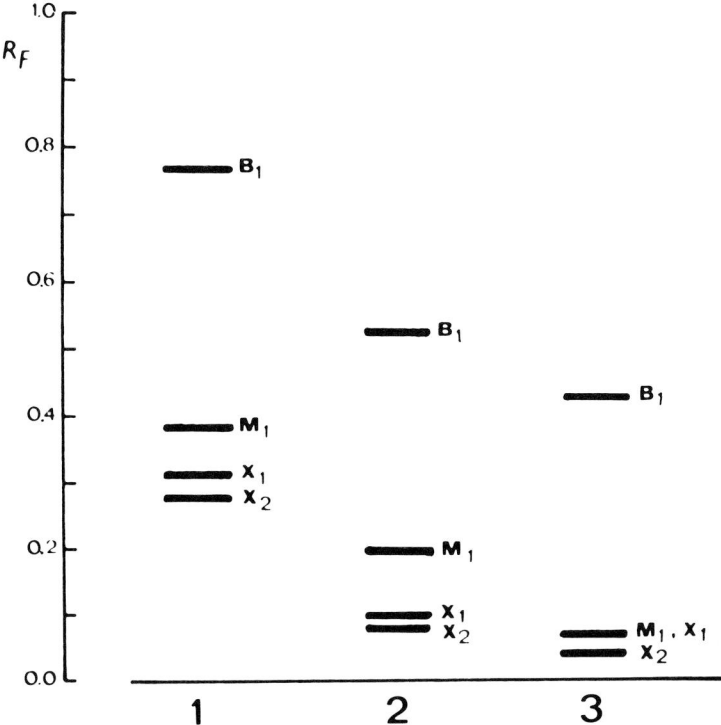

FIGURE 52. R_F values of aflatoxin B_1, M_1, and metabolites X_1, X_2.

children with RS. The character of these muscle changes was similar to that of Zenker's degeneration, in which an excessive accumulation of lactic acid is the assumed pathogenic mechanism.[18] It was suggested that a similar pathogenic mechanism could lead in this syndrome to changes in the striated muscles, in which the metabolic pathway of the Krebs' cycle is impaired.

The etiology of this disorder in newborns is unclear. Viral infection, inborn metabolic disorders, toxins, or a combination of these factors have been presumed.[16,17] An attempt at virus isolation in two cases was negative. the possibility of a genetic metabolic disorder could not be excluded in one of the familial cases. Analogous cases of a familial fatal steatosis in newborn infants, bearing some striking pathological resemblance to these cases, were previously reported by Peremans et al.[19] and Satran et al.[20]

The patient reported by Peremans et al.[19] was the 14th infant of parents who were cousins. All pregnancies were uneventful, and the infants were normal at birth. Five of them, all boys, died within 48 h of birth and one girl died at the age of 2 years. In two cases, pallor of the liver was seen at the autopsy. The 14th infant was well after birth and died on the sixth d of life. On the second day, the infant became hypotonic and lethargic. The blood-sugar level fell to 12 mg/100 ml. A histological study showed generalized fatty degeneration of the viscera. The authors assumed that this was a disorder which was genetically transmitted in a recessive manner. Satran et al.[20] reported three infants from the same family who died from the fatty-liver disease and considered the possibility of RS, but because of the normal blood and spinal fluid sugar and no CNS findings, this diagnosis was discarded.

Aflatoxin demonstrated in the liver samples of two of the three newborns could be an important factor in the etiology of this disorder. It is known that this mycotoxin produces Reye-like morphological changes in experimental animals, and it has been demonstrated in organs of the older children with RS.

Hypothetically, there were two possibilities of exposure to aflatoxin in the neonatal period: either through breast feeding of the mother's contaminated milk or transplacentally. In animal experiments, it has been shown that AFB_1 passes the placental barrier, and it has been detected in the liver of fetuses.[21] The toxin causes fetal death and fetal growth retardation. Das et al.[22] have reported a high incidence of neonatal death within 2 d post-partum in offsprings of the rats receiving AFB_1 in late pregnancy.

Due to aflatoxin toxicity, the neonatal fatalities were associated with pulmonary insufficiency of surfactant lipids. The authors have shown that AFB_1 causes a decreased synthesis of phospholipids and lecithin, the components being responsible for alveolar stability which is essential for the survival of newborns. Even though the phospholipid synthesis was reduced, there was an increase in the synthesis of total lipids, attributable to the increase in the synthesis of neutral lipids. A similar type of increase in the synthesis of neutral lipids has been shown to occur in tissues due to toxicity.

This experimental study supports the hypothesis that intrauterine exposure to aflatoxin could be involved in the pathogenesis of this disorder in newborns. The presence of hyaline membranes in the lungs of newborns, in addition to a fatty change of the viscera, may provide evidence that the synthesis of surfactant lipids was inhibited, while the synthesis of neutral lipids was similarly increased as reported in the experiment.[22] The hypothesis concerning a possible role of aflatoxin in the pathogenesis of the neonatal disorder is further supported by the following:

1. All cases but one were observed in a single geographic region in 1972.
2. Except for one case, all the infants came from families living in the country.
3. The mothers of all but one worked on cooperative farms, where contact with aflatoxin (contaminated agricultural products) is more probable than in a town or city.

Although it seems that RS in a newborn is quite exceptional, it is likely that similar cases exist but escape attention.

VI. EXPERIMENTAL MODELS OF REYE'S SYNDROME

Epidemiological surveys have suggested the possibility that environmental factors may be contributory to RS, acting synergistically with a viral infection. A number of environmental toxins, such as herbicides, insecticides, and surfactants, have been studied in various animal models and many of the histopathological and biochemical abnormalities found in RS have been replicated in these diverse animal models.[23-26] One of the exogenous toxins considered to be involved in the etiology of RS is aflatoxin, which has been used in the following experiment.

A. AFLATOXIN B_1 AND INFLUENZA A VIRUS INTERACTION IN MICE
1. Materials and Methods

Sixty 7-week-old white male mice (outbred strain ICR) were divided into six groups of ten animals each and housed in a constant–temperature (25°C) room. Water and feed were supplied *ad libitum*.

Chemicals — Aflatoxin B_1 (Spofa, Czechoslovakia), dissolved in DMSO, was administered orally in a single dose of 4 mg/kg of body weight (= $1/2$ of LD_{50} for mice), and 2 mg/kg of body weight (= $1/4$ of LD_{50} for mice).

Virus — Influenza A virus strain S/N, a recombinant of NWS (A-H_0N_1), and Singapore 57 (A-H_2N_2), was passaged on suckling mice, the allantois of a chicken embryo and diploid cells (LEP human embryonal lung cells, USOL, Prague). The animals were inoculated intranasally with 0.05 ml of the concentrated fluid of influenza virus (titer 10^8 ml) 12 h after their exposure to aflatoxin. Treatment of the individual groups of animals is presented in Table 32. The animals were killed on day 4, 5, 6, 7, and 8 after virus inoculation. The organs were taken for a histological investigation, fixed in 10% formalin, and stained with hematoxylin and eosin and Sudan.

TABLE 32
Mouse Model for Reye's Syndrome
Treatment Protocol

Group I	AFB_1 4 mg/kg of body weight
Group II	AFB_1 2 mg/kg of body weight
Group III	0.05 ml of influenza A virus (titer 10^8/ml)
Group IV	AFB_1 4 mg/kg of body weight + 0.05 ml virus
Group V	AFB_1 2 mg/kg of body weight + 0.05 ml virus
Group VI	DMSO

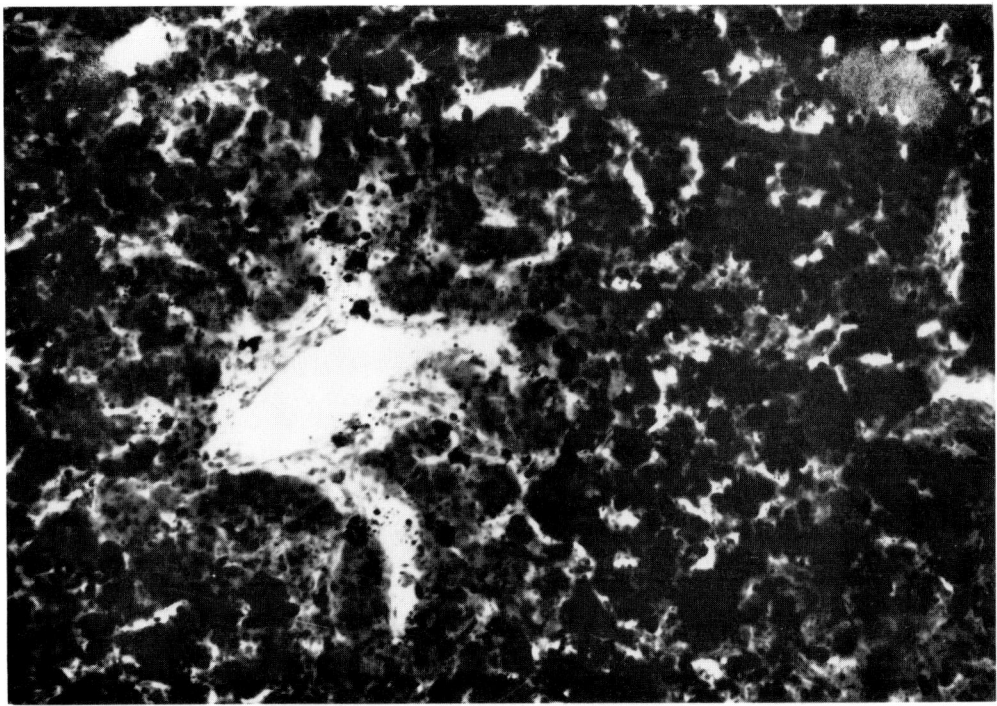

FIGURE 53. Diffuse fatty change of the liver of mouse on the fifth day after aflatoxin administration and virus inoculation (Sudan, ×132).

2. Results

In the first two groups of the animals treated with AFB_1 alone, a mild degree of steatosis in the liver at the periphery of the lobules was found on day 4, 5, and 6. No cell necrosis was present. The histological sections of the heart and kidney revealed no pathological changes. In the third group of the animals innoculated with the virus alone, only a discrete peribronchial mononuclear infiltration in the lungs was found. The other organs were without pathological changes. Marked morphological changes were present on the day 5 in the fourth group of the animals treated simultaneously with AFB_1 (4 mg/kg) and the virus. A diffuse fatty degeneration and a mild activation of Kupffer cells were noted in the liver (Figure 53). A marked steatosis of the tubular kidney cells and the myocardium was also present (Figures 54, 55). In the fifth group of animals which received 2 mg/kg of AFB_1 and the virus, only steatosis of a milder degree was present in the liver. The other organs were without pathological changes. Table 33 indicates the degree of fatty change in the individual groups of the animals. No pathological changes were found in the control group treated with DMSO.

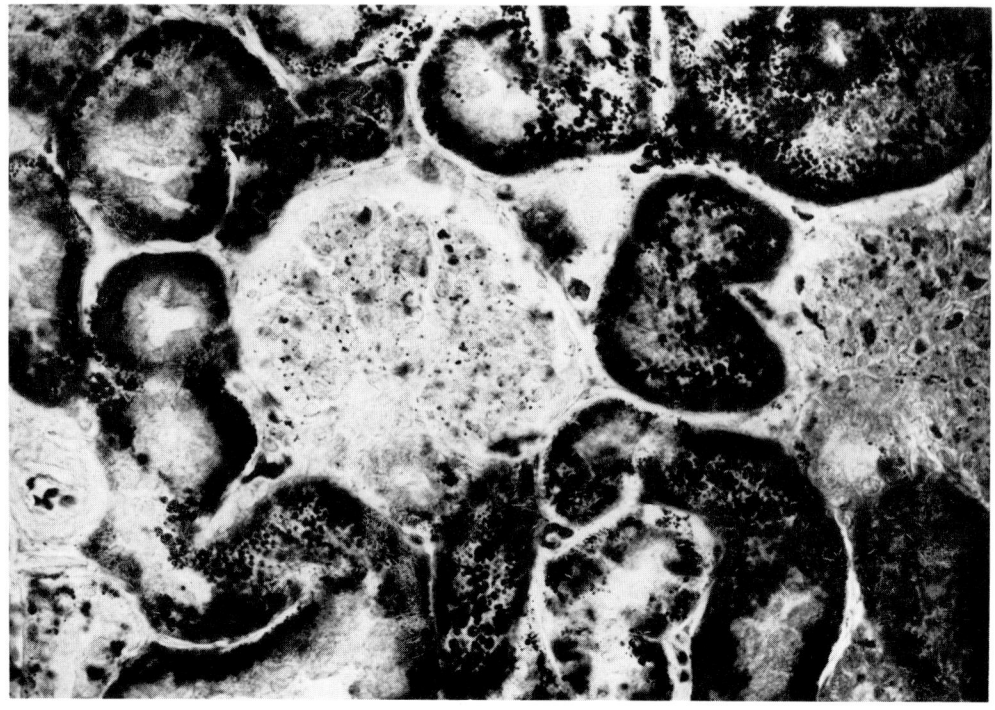

FIGURE 54. Marked steatosis of the proximal tubules of the kidney in mice on the fifth day after aflatoxin and virus treatment (Sudan, ×220).

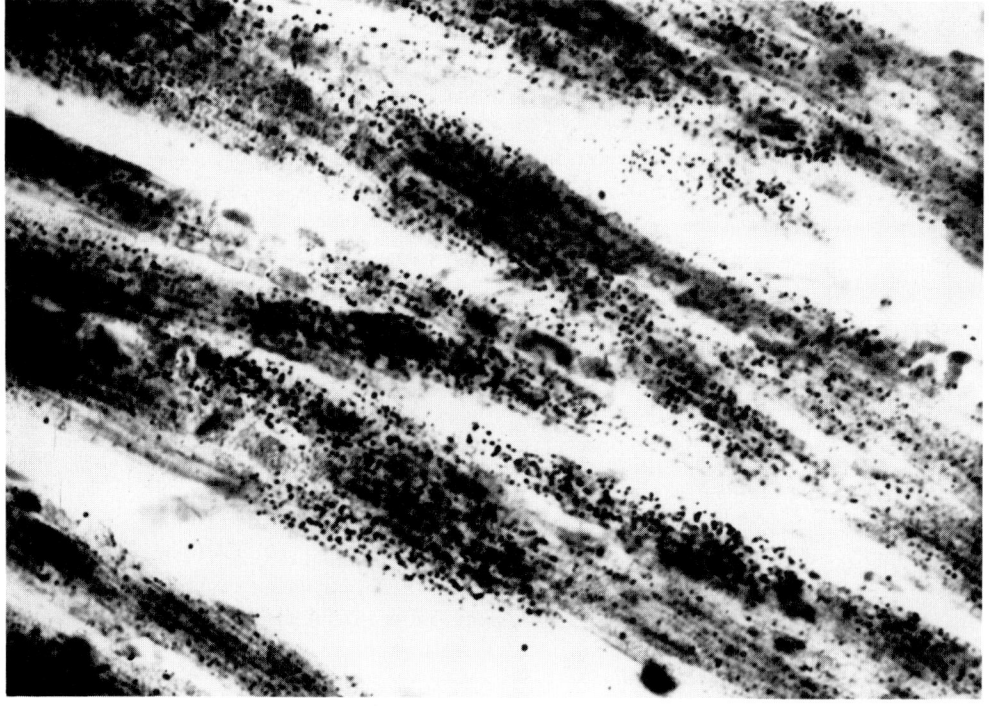

FIGURE 55. Fine droplet steatosis of the myocardium in mice given aflatoxin and virus (Sudan, ×220).

TABLE 33
Mouse Model for Reye's Syndrome — Degree of Fatty Change in Mice Organs

	Group	Organs	Day 4	Day 5	Day 6	Day 7	Day 8
I.	AFB_1 (4 mg/kg)	Liver	++	++	+		
		Heart					
		Kidney					
II.	AFB_1 (2 mg/kg)	Liver	++	+			
		Heart					
		Kidney					
III.	Virus 0.05 ml	Liver					
		Heart					
		Kidney					
IV.	AFB_1 (4 mg/kg)	Liver	++	+++	++		
	+ virus 0.05 ml	Heart	+	+++			
		Kidney	+	+++			
V.	AFB_1 (2 mg/kg)	Liver	++	++			
	+ virus 0.05 ml	Heart					
		Kidney					
VI.	DMSO	Liver					
		Heart					
		Kidney					

Morphological changes similar to those seen in RS (fatty degeneration of the liver, kidney, and myocardium) were present in the group of animals treated with a higher but nonlethal dose of AFB_1 on day 5 after virus inoculation, whereas the animals treated with AFB_1 alone revealed only a mild fatty change of the liver. No fatty change was found in the animals treated with the virus alone.

It seems that a viral infection appears to potentiate the effect of aflatoxin. The enhancing effect of the virus and aflatoxin was already earlier reported by Svoboda et al.[27] in marmosets given AFB_1 and the Barker strain of hepatitis virus.

It was of particular interest that a marked fatty change of the organs was found on day 5 after virus inoculation, whereas on the following days only a mild or no fatty change was present. Similar findings are known from the liver biopsies of RS patients, in whom a diffuse fatty change is present only in the acute phase of the illness and disappears after a short time in the surviving patients.[28] The mechanism of the interaction among the virus, toxin, and host is not yet defined. There are some principal questions; it is a direct synergistic effect of both the agents, or can it be that either the viral infection allows a release of the stored toxin or the depressed host immunity allows an increased viral replication? All these questions remain unanswered at present, and further work directed to this problem is needed.

B. ULTRASTRUCTURAL CHANGES IN INTERACTION OF ADENOVIRUS 3 AND AFLATOXIN B_1 ON TISSUE CULTURES OF HeLa CELLS

The simultaneous effect of both the virus and aflatoxin has been studied *in vitro* on HeLa cell cultures.

1. Materials and Methods

Adenovirus type 3 was used for the experiment because it had been isolated from the liver of several patients with RS. (The virus was provided by Drs. Sirucek and Brucková from the Virological Institute in Prague.)

Pure AFB_1 (Calbiochem, California) was dissolved in minimal essential medium (MEM) at concentrations of 0.2, 2.0, and 20 µg/ml of the medium. The concentrations were measured immediately before the experiment.

The pure media in 2-d-old cultures of HeLa cells were replaced by the media containing AFB_1 at concentrations of 0.2, 2.0, and 20 µg/ml. After 1 h of incubation at 37°C, the tissue cultures were inoculated with an infectious medium containing 10^6 viral particles per milliliter, and the virus was allowed to absorb for 2 h. The cells were then cultured in the media with aflatoxin until the end of the experiment. As controls, the tissue cultures of HeLa cells inoculated with adenovirus 3 and cultivated in the medium without aflatoxin were used. After 72 h HeLa cells from both the cultures with aflatoxin and the controls were trypsinized, washed in a balanced salt solution, and centrifuged to form small pellets. The pellets were then fixed in 3% glutaraldehyde, post-fixed in 1% osmium tetroxide, dehydrated in ethanol, and embedded in Durcupan. The sections cut on an EM microtome were stained with uranyl acetate and lead citrate, and viewed in a Tesla BS 500 electron microscope. Digestion by pronase was carried out on ultrathin sections which had been treated with 10% H_2O_2 at 40°C for 10 min, and later with 0.01% pronase at 40°C for 30 min.

2. Results

HeLa cells cultivated in the medium without aflatoxin 72 h after the infection with adenovirus 3 showed intranuclear masses of dense osmiophilic material and crystals composed of viral particles. Many of the viral particles had a dense central body enclosed with a sharply defined membrane. The virus was apparently formed in connection with and at the expense of the dense material. In addition, intranuclear tubules composed of laminated membranes were found in the nuclei of some cells (Figure 56). The cytoplasm of the cells showed various stages of degenerative changes which resulted in its total vacualization and a rupture of cell membranes.

HeLa cells cultivated in the medium with AFB_1 at concentrations of 0.2 and 2.0 µg/ml of the medium 72 h after the inoculation with adenovirus 3 revealed a disintegration of crystals composed of viruses, which resulted in viral particles being mostly irregularly scattered on the periphery of nuclei (Figure 57). The number of viral particles with a dense central body, in comparison with empty viral capsides, was nearly the same as in the cells cultivated in the medium without aflatoxin. Clusters of dark osmiophilic material did not show any changes in comparision with the controls. HeLa cells cultivated in the medium containing AFB_1 at a concentration of 20 µg/ml, 72 h after virus inoculation, contained nuclei showing only clusters of dark osmiophilic material, and numerous intranuclear tubules originated probably from the clusters of dark osmiophilic material (Figures 58, 59). Complete viral particles were very rare. A small number of the nuclei contained free viral particles scattered in the nucleus. Crystalline structures were not present. The majority of particles were viral capsides without a central body (Figure 60). In these cells, empty viral capsides were often in intimate contact with intranuclear tubules (Figure 61). There was a striking similarity between the structural capsides and the individual lamellae of intranuclear tubules (Figure 62). After digestion by pronase, the clusters of dark osmiophilic material, intranuclear tubules, and viral capsides disappeared (Figure 63).

Observation of the replication of adenovirus 3 in nuclei of the HeLa cells cultured in the medium without aflatoxin corresponded to the findings of Morgan et al.[29] and Block et al.[30] These authors described the formation of viral particles from clusters of dark osmiophilic material in the nuclei of HeLa cells, the existence of aggregates of these viruses arranged as crystalloids, and the existence of intranuclear tubules, which are specific to adenovirus 3. The disintegration of viral crystalloid aggregates was observed more frequently in the HeLa cells cultured with AFB_1 (0.2 and 2.0 µg/ml) than in the control HeLa cells cultivated without aflatoxin. This disintegration of viral crystalloid aggregates usually occurred at the final stage of the evolutionary cycle of adenoviruses before disintegration of the nuclear membrane.[1,2] Therefore it seems that the presence of AFB_1 in the medium accelerated the final phase of the evolutionary cycle of adenoviruses. The most prominent changes were found in the HeLa cells

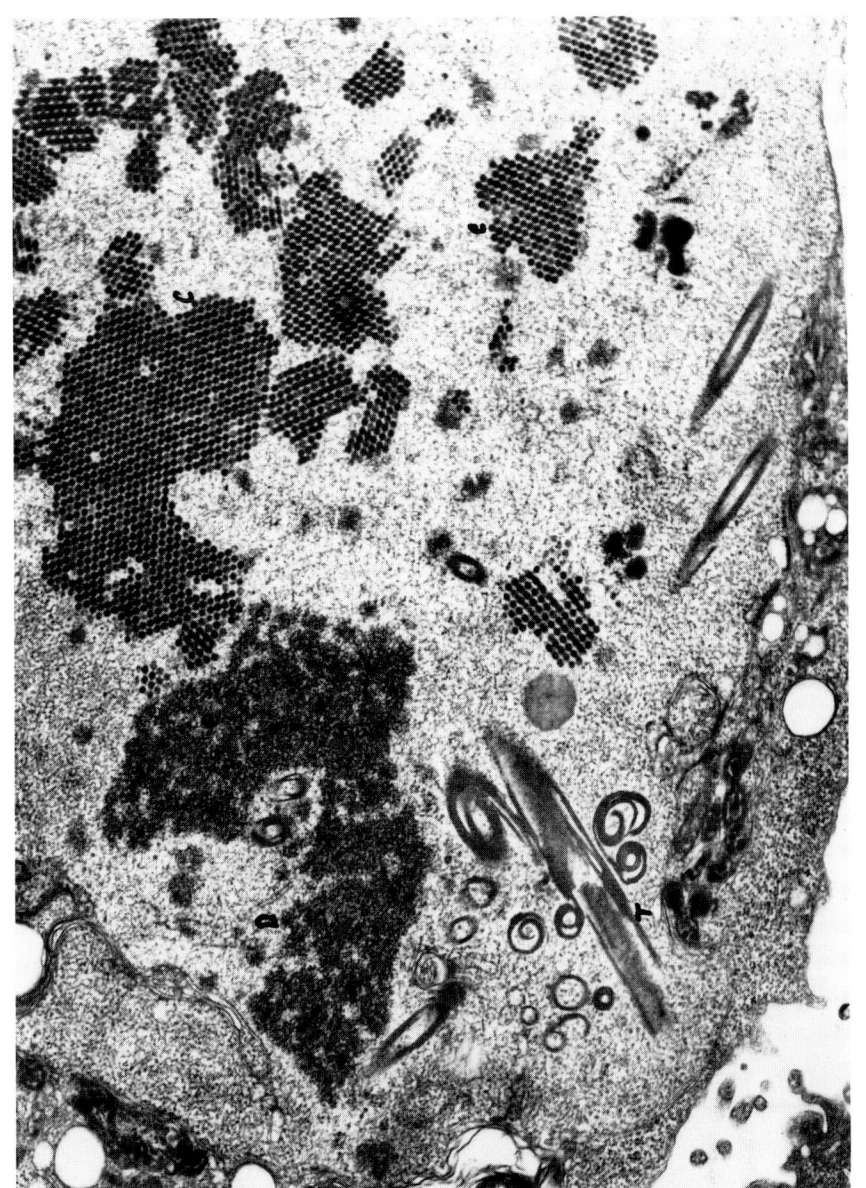

FIGURE 56. Adenovirus 3 — infected HeLa cells in medium without aflatoxin B_1. In the nuclei are masses of a dense osmiophilic material (D), crystals composed of viral particles (C), and intranuclear tubules (T) (×12,040).

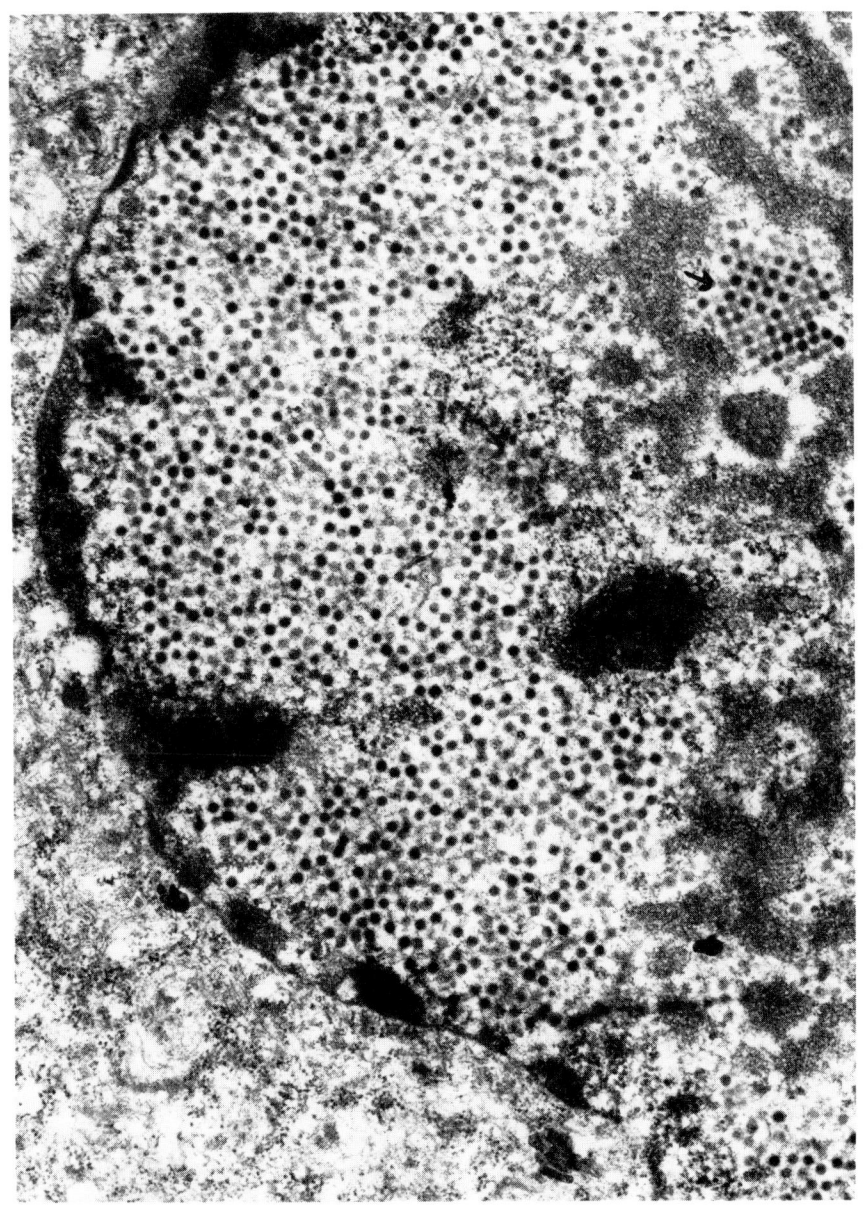

FIGURE 57. Adenovirus-infected HeLa cells in medium containing aflatoxin B_1 at the concentration of 0.2 μg/ml of the medium. Free viral particles scattered in the nuclei. Crystallic structures are very rare (arrow. ×20,640).

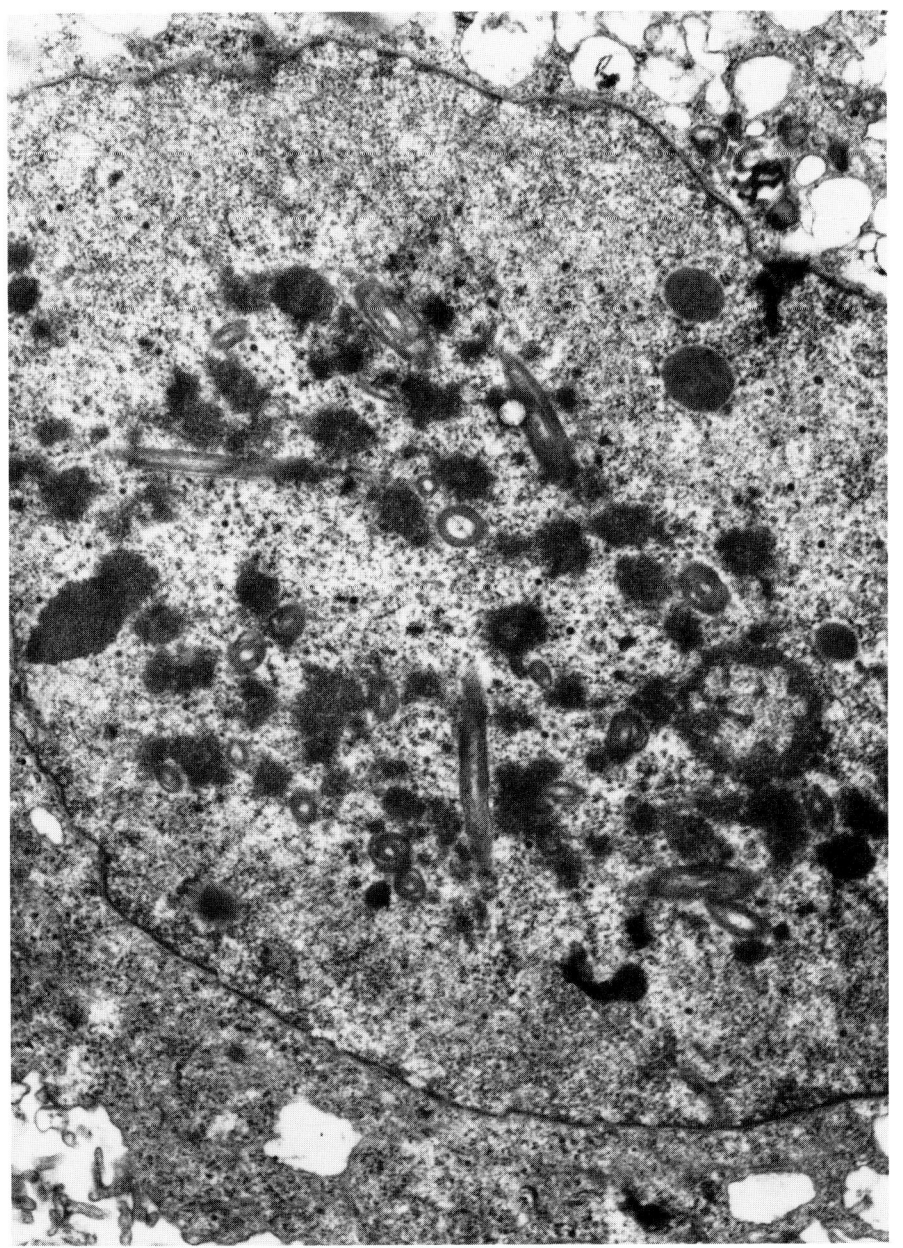

FIGURE 58. Adenovirus 3 — infected HeLa cells in medium containing aflatoxin B_1 at the concentration of 20 µg/ml of the medium. The nuclei contain clusters of a dark osmiophilic material and intranuclear tubules. Viral particles are scarcely present (×11,610).

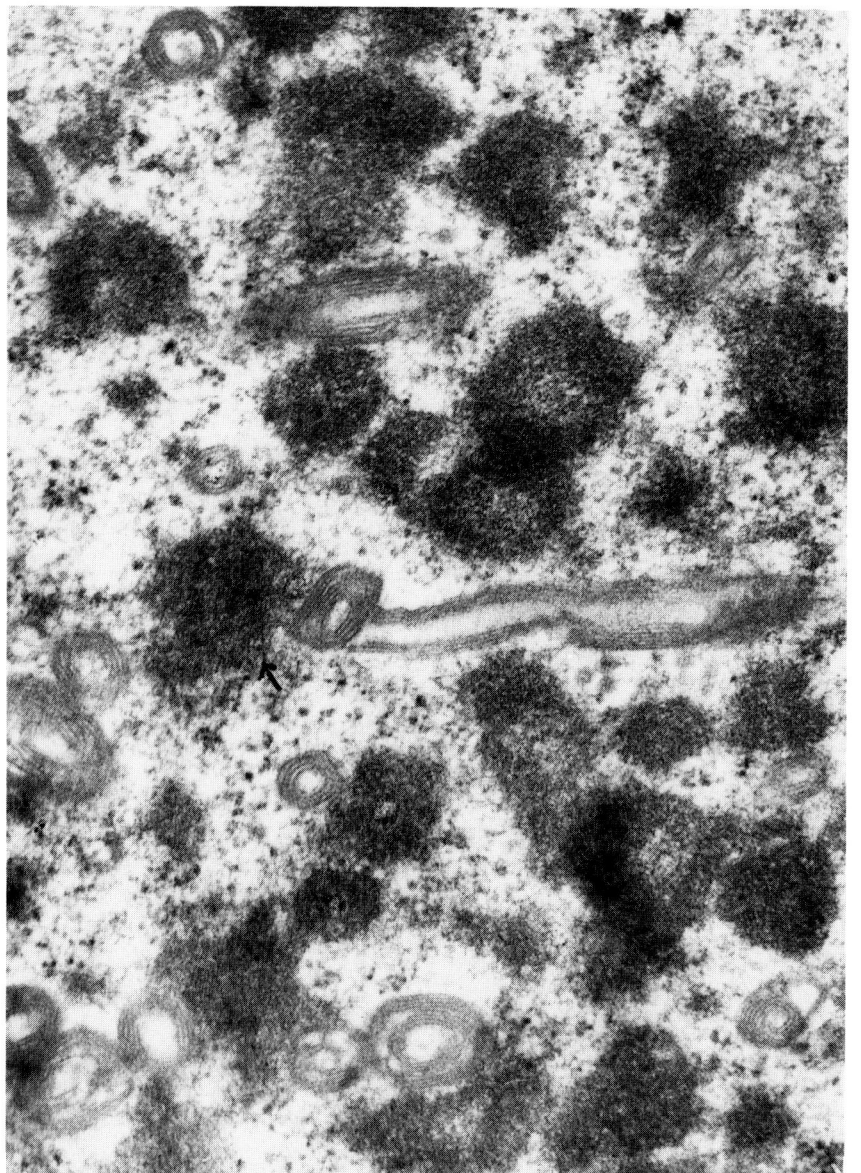

FIGURE 59. Intranuclear tubules probably originate from the clusters of dark osmiophilic material (arrow). Aflatoxin B_1 concentration 20 µg/ml of the medium (×25,800).

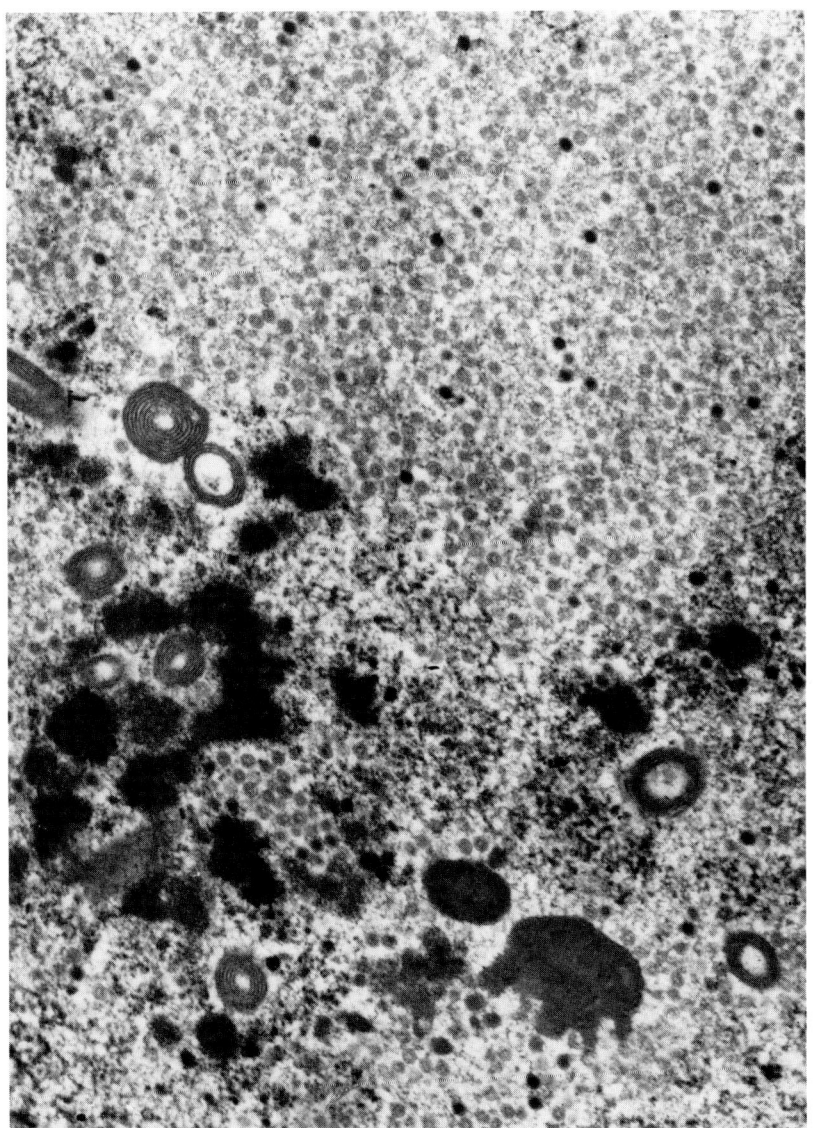

FIGURE 60. The nuclei contain a great amount of viral capsides without central bodies. The complete viral particles seldom occur. Intranuclear tubules (T) are present. Aflatoxin B_1 concentration of 20 μg/ml of the medium (×20,880).

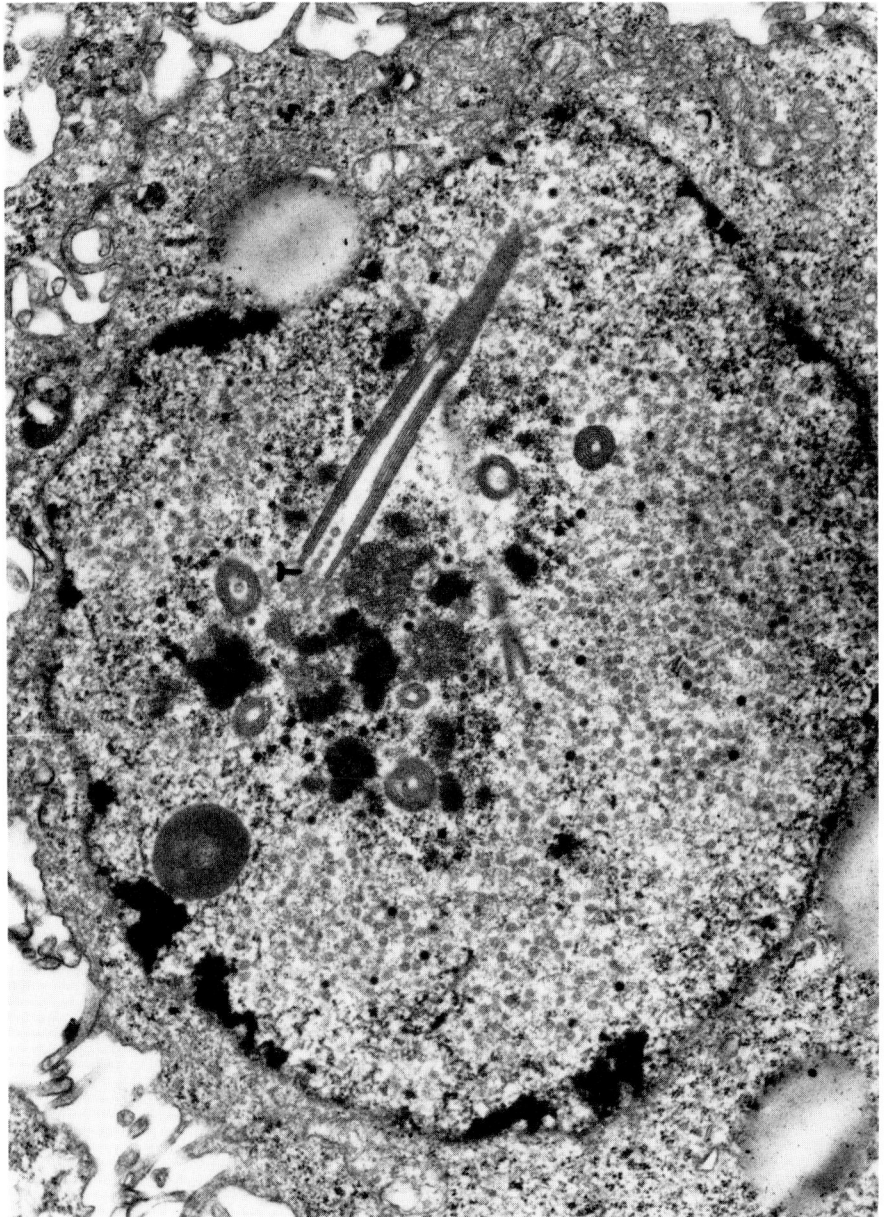

FIGURE 61. Numerous viral capsides are in the nucleus, some of which are in the lumina of the intranuclear tubules (T). Aflatoxin B_1 at the concentration of 20 μg/ml of the medium (×14,620).

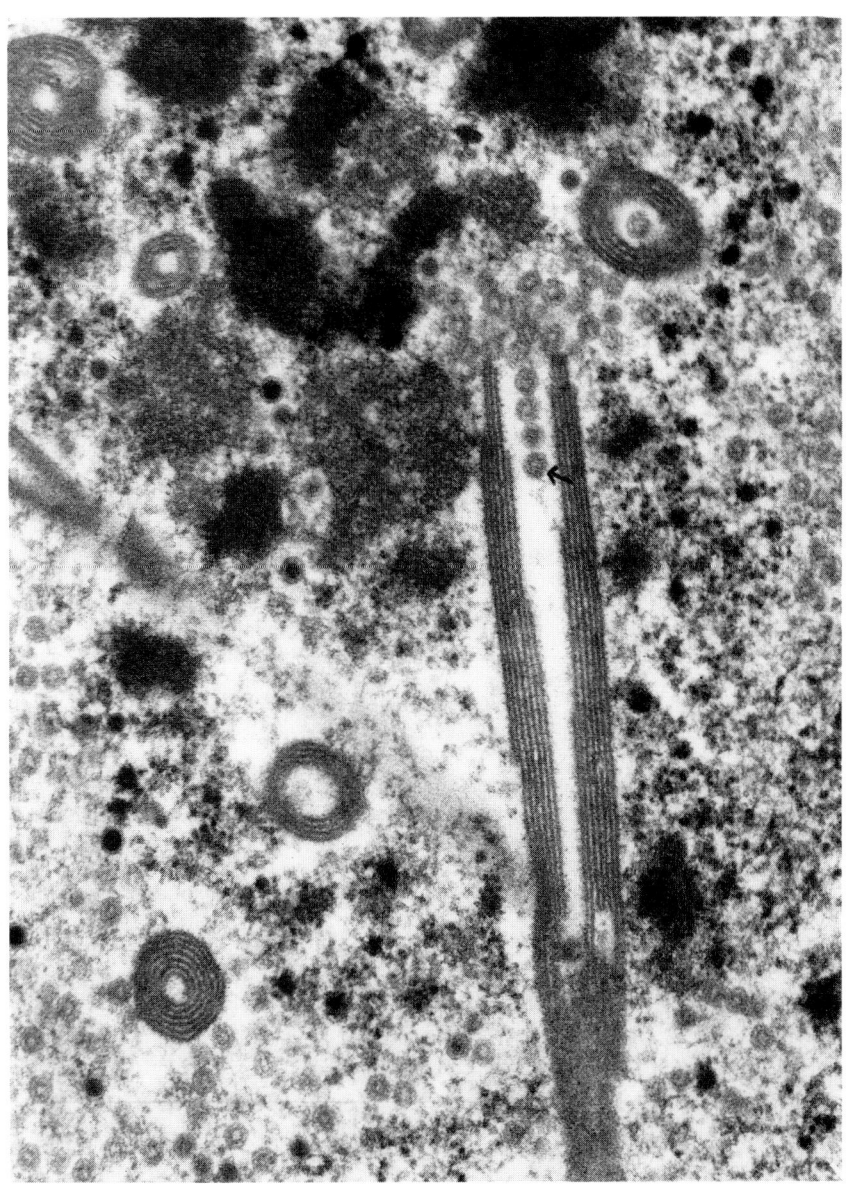

FIGURE 62. Viral capsides in the lumina of intranuclear tubules (arrow). Detail of Figure 61 (×34,400).

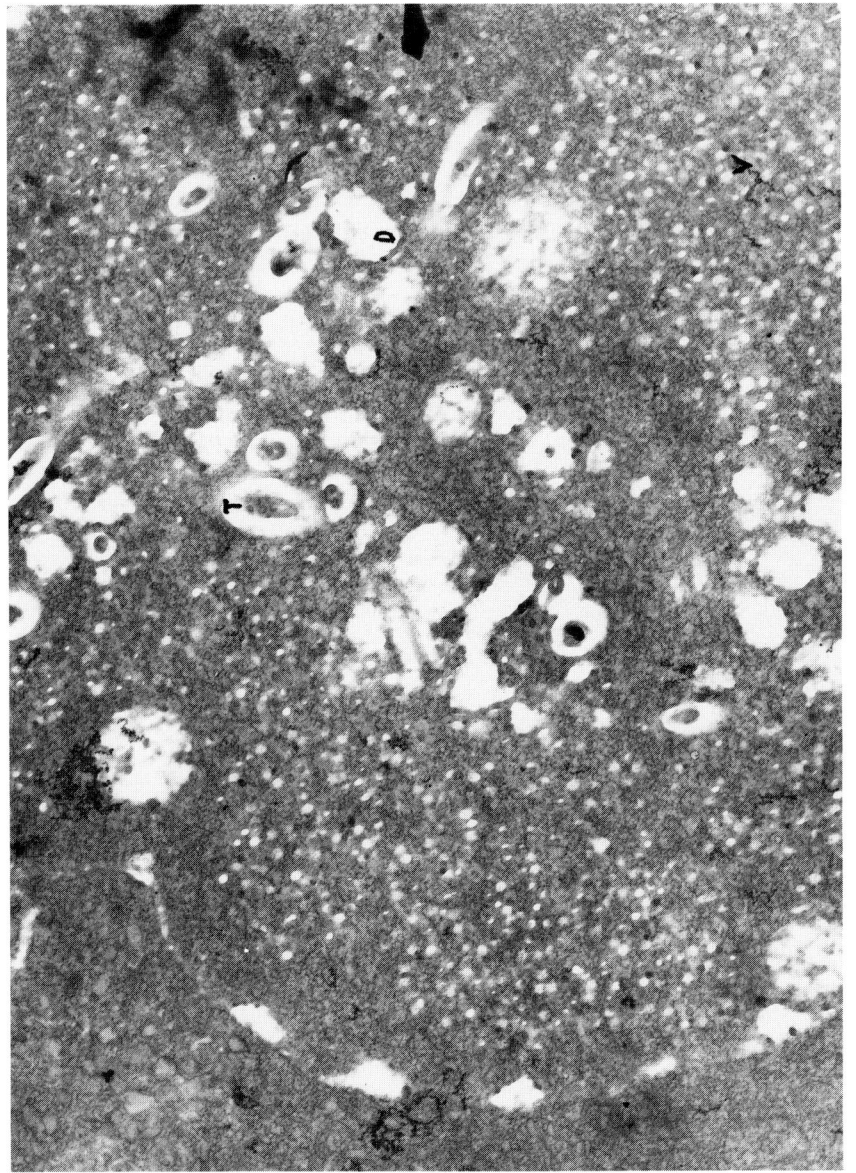

FIGURE 63. The nucleus digested by pronase. Dark osmiophilic material (D), intranuclear tubules (T), and viral capsides (V) have disappeared (×14,620).

cultivated in the medium with AFB_1 at a concentration of 20 µg/ml of the medium. Nuclei of the cells showed abundant clusters of osmiophilic material, intranuclear tubules and empty viral capsides. All those structures could be removed by digestion with pronase, which indicated a protein character of all these structures. In contrast, complete viral particles containing DNA were very rare. Thus it has been suggested that AFB_1 inhibits only the viral DNA synthesis, but the synthesis of viral proteins is not affected. These findings are in agreement with the observation of Crook et al.,[31] who has demonstrated that AFB_1 inhibits a synthesis in vaccinia-infected HeLa cells, but without the involvement of proteosynthesis (DNA-polymerase).

On the basis of these results, the following hypothesis has been suggested: certain concentrations of AFB_1 can induce a formation of defective viruses, but a viral proteosynthesis as the carrier of specific antigenic properties is not inhibited. It is known that a positive isolation of viruses in patients with RS is exceptional, although serological investigations have proved increased titers of antibodies against various viruses in many of these cases. Therefore it may be suggested that aflatoxin can inhibit the normal development of viruses, which could be the cause of the unsuccessful isolation of viruses.

In addition, it has been supposed that the viral antigen can interact with the toxin to produce severe cellular abnormalities. The clinical, biochemical, and pathological features, similar to those in RS due to the viral antigen without virus replication, have been reported in animal models[32] as well as in children with a disease simulating RS.[33]

C. COMMENTS AND CONCLUSION

RS was suggested to be the result of several interrelated factors.[34,35] Reye[36] in his original paper could find no constant etiological agent and concluded that he was not convinced that etiology was identical in every case.

Although RS was originally associated particularly with viral infections,[37,38] additional data suggested that environmental toxins may also be important in the etiology of this illness.[23-25] The discovery of aflatoxin in 60% of our study subjects is of particular importance and may demonstrate a real link between this toxin and some cases of RS.

What are the facts supporting this presumption?

1. Most of the affected children were infants under 1 year of age. This age distribution is consistent with the fact that in all species of animals that have been studied, aflatoxin sensitivity is the greatest in the young.[39]
2. The clinicomorphological features, similar to those seen in RS, developed in young monkeys fed aflatoxin B_1.[40] Clinical symptoms of the monkeys included cough, vomiting, diarrhea, and coma, accompanied with hypoglycemia, elevated free fatty acids, and transaminases. The post-mortem examination revealed cerebral edema with neuronal degeneration, fatty change of the liver, kidney, myocardium, and lymphocytolysis in the lymphatic tissue, and lesions almost identical to those found in RS. The only difference was presented by hepatic necrosis and bile-duct proliferation, changes which are uncommon in RS cases ending in death within several hours or days, but which were found in the liver of children with a protracted course of the illness.[13,41] The ultrastructural changes involving swollen mitochondria with loss of dense bodies, glycogen depletion, and lipid vacuoles in the cytoplasm of the hepatocytes and tubular cells of the kidney, found in our cases, were consistent with those described in RS.[42] In addition to the mitochondrial abnormalities, nucleolar alterations were found in both the livers and kidneys of children in whom aflatoxin was chemically detected. Enlarged nuclei with interchromatine granules, disorganization of chromatin, and zoning of nucleoli paralleled the changes reported in animals treated with AFB_1.[7-9,43]
3. Numerous metabolic abnormalities are characteristic of RS. The most constant biochemical disorders are prolonged prothrombin time, elevated transaminases, free fatty acids,

hyperammonemia, and hypoglycemia. Similar biochemical abnormalities have been reported in animals treated with AFB_1. Prolonged prothrombin time and reduced activity of coagulation factors, synthesized in the liver (fibrinogen, prothrombin, factors V, VII, IX, and X) have been reported in RS[44] as well as in aflatoxin-fed animals.[45,46] Elevated serum transaminases and free fatty acids — characteristic signs of RS — were found elevated in animal experiments after AFB_1 administration.[40,43,47] Decreased activity of the mitochondrial liver enzymes — ornithine transcarbamylase and carbamylphosphate synthetase — is known to be present in RS. Recently, it has been shown that AFB_1 depresses the mitochondrial enzymes — carbamylphosphate synthetase and ornithine transcarbamylase — without affecting the cytosolic urea cycle enzyme arginase.[48] Hypoglycemia, often found particularly in infants with RS, was observed in monkeys treated with AFB_1.[40] Altered lipid metabolism and lipid transport have been reported in experimental aflatoxicosis[49] as well as in RS.[50] A general diminution of glycogen in patients with RS has been well documented, and a depletion of glycogen by aflatoxicity has also been reported in rats[7] and swine.[51] Hypertyraminemia is a common metabolic abnormality in RS primarily caused by impairment of hepatic mitochondrial monoamine oxidase,[52] correlating with the stage and duration of the coma. Accumulation of tyramine and its product, octopamine, in the brains of RS patients has been reported;[53,54] tyramine-induced neurotoxicity has also been well documented on an animal model.[55] Thus it has been concluded that the encephalopathy of RS may be linked to the presence of false transmitters in the brain. Recent studies have demonstrated that AFB_1 also possesses a neurotoxic effect resulting in a degeneration of the central and peripheral nervous systems.[56] This neurotoxic effect of AFB_1 appears to be specific to dopaminergic pathways, possibly by selectively perturbing the conversion of tyrosine into biogenic catecholamine neurotransmitters.[57] Because RS is a disease characterized by neuronal degeneration, it has been assumed that aflatoxin may indeed be involved in the etiology of this illness.[57]

The concentrations of AFB_1 found in the liver of children who had died of RS varied from 5 to 1760 μg/kg, with an average between 80 to 200 μg/kg of liver weight. In spite of the differences in the amount of the toxin, there were no differences in the morphological findings in the liver, where only a diffuse fatty change without any necrosis was present. In this way, RS differs from the findings in patients during the outbreak of aflatoxicosis reported in India[58] and Kenya,[59] in whom clinical symptoms such as abdominal discomfort, anorexia, malaise, and jaundice as well as the morphological findings of hepatic necrosis corresponded to aflatoxin poisoning known from the animal experiment. This supports the hypothesis that aflatoxin in RS may act only as a contributing factor, most probably in association with viral infection, and therefore, RS could not be regarded as a "toxicosis" in the strict sense.

The interaction of viruses and toxins has been well documented.[24,25,60] Morphological changes corresponding to those seen in RS were found in mice treated with influenza A virus and a nonlethal dose of AFB_1. A recent study on beagle pups treated with a live measles vaccine, aspirin, and AFB_1 has shown that criteria of increased serum transaminase activity, decreased blood glucose, hyperammonemia, a fatty infiltration of the liver, and hepatic mitochondrial alterations were produced more consistently in puppies treated with AFB_1 and the virus.[61]

The exact mechanism of virus-toxin interaction is not known. It has been suggested that either both virus and toxin act directly or indirectly at a common cellular locus,[23] or the toxin impairs the normal immune defense.[62] Impairment of the immune defense has recently been assumed in RS. The morphological changes found in the lymphatic organs[2,63,64] as well as the ultrastructural alterations[65-67] seem to support this hypothesis. Aflatoxin has been shown to impair the immune system in many ways, including the effects on antibody formation,[68,69] complement,[70] phagocytosis,[71] and the production of interferon.[72] Therefore, it seems plausible that rather the immunosuppressive than direct effect of aflatoxin may be important in the pathogenesis of RS.

5. A nationwide surveillance on RS has never been realized in Czechoslovakia, although the disease has been recognized in this country since the 1960s. So epidemiology does not illustrate the real situation of RS in the whole children's population in Czechoslovakia. Nevertheless, some general comments can be made from the data collected from the case reports of children who had died and were autopsied. The findings in a group of 137 studied cases (1958 to 1986) indicated that the disease occurred sporadically throughout the year, particularly involving children under 1 year of age (61%), and was more often in children from rural regions (62%) compared to those from urban areas (37%). The most frequent prodromal symptoms were upper respiratory infections, found in 83% of the cases. Out of 51 attempts to isolate influenza, herpes, coxsackie, and adenovirus, only four attempts were successful and adenovirus types 2, 3, and 5 were recovered from patients' tissues. Serological investigations positively revealed influenza A antibodies in 7 out of 13 and hepatitis A antibodies in 2 out of 5 investigated cases. The patients' contact with viral infection in the family or children's groups proved various infections such as infectious hepatitis, a flu-like disease, adenovirus infection, morbilli, and varicella. Salicylate therapy was used in 29% of cases with known therapy. The epidemiological findings of our study differed from those reported by CDC because of a high proportion of patients under 1 year of age, and the absence of a marked winter peak associated with the outbreak of influenza epidemics, as has been demonstrated by North American studies.[73,74] Data of the National Influenza Surveillance in Czechoslovakia indicate that moderate influenza outbreaks, caused mainly by influenza A (H_1N_1), have occurred each year since 1974, December through March, with two major epidemics in 1979/80 and 1981/82 caused by influenza A (H_3N_2) and influenza B in minor extent, but no clustering of RS was noticed. The cause of this difference is not clear. It cannot be excluded that the series is too small so far to demonstrate such an association, and furthermore, only 37% of the cases were examined virologically. Although the prodromal symptoms as well as epidemiological data have provided evidence that various viral infections were associated with RS cases, a sporadic occurrence, lack of family epidemics, and lack of clustering of patients within the patient's groups have indicated that other factors may be involved in the etiology of RS.

Salicylates, which have been suspected to be a contributing factor in the etiology of RS, were not usually used in this group of patients.

6. The presence of aflatoxin found in the livers in 60% of RS cases and in the food consumed by the patients shortly before the onset of the disease provides further convincing evidence for aflatoxin to be one of the etiological factors in the development of RS. The results of this follow-up study have attracted the attention of government institutions in Czechoslovakia which have adopted legislative arrangements involving a current control and investigation of food and food substrates, particularly of those intended for children. To avoid possible contamination in the households, each packet of powdered milk and infant formula has been provided with a label advising to keep the food dry and cold and consume it within 14 d after opening.

REFERENCES

1. **Dvořáčková, I., Vortel, V., and Hroch, M.,** Enzephalitisches Syndrom mit Leberverfettung, *Zbl. Pathol. Bakteriol.,* 106, 573, 1964.
2. **Bourgeois, C. H., Olson, L., Comer, D., Evans, H., Keschamras, N., Cotton, R., Grossman, R., and Smith, T.,** Encephalopathy and fatty degeneration of the viscera: a clinico-pathologic analysis of 40 cases, *Am. J. Clin. Pathol.,* 56, 558, 1971.
3. **Olson, L. C., Bourgeois, C. H., Cotton, R. B., Harikul, S., Grossman, R. A., and Smith, T. J.,** Encephalopathy and fatty degeneration of the viscera in Northeastern Thailand. Clinical syndrome and epidemiology, *Pediatrics,* 47, 707, 1971.
4. **Stubblefield, R. D. and Shannon, G. M.,** Aflatoxin M_1: analysis in dairy products and distribution in dairy foods made from artificially contaminated milk, *J. Assoc. Off. Anal. Chem.,* 57, 847, 1974.
5. **Simard, R. and Bernhard, W.,** Le phénomene de la ségrétation nucléolaire: spécificité d'action de certains antimétabolites, *Int. J. Cancer,* 1, 463, 1966.
6. **Bernhard, W., Frayssinet, C., Lafarge, C., and LeBreton, E.,** Lésions nucléolaires précoces provoquées par l'aflatoxine dans les cellules hépatiques du rat, *C.R. Acad. Sci (Paris),* 261, 1785, 1965.
7. **Svoboda, D. J., Grady, H. J., and Higginson, J.,** Aflatoxin B_1 injury in rat and monkey liver, *Am. J. Pathol.,* 49, 1023, 1966.
8. **Bauer, L., Tulusan, A. H., and Muller, E.,** Ultrastructural changes produced by the carcinogen aflatoxin B_1 in different tissues, *Virchows Arch. B:,* 10, 275, 1972.
9. **Svoboda, D. J. and Higginson, J.,** A comparison of ultrastructural changes in rat liver due to chemical carcinogens, *Cancer Res.,* 28, 1703, 1968.
10. **Davidson, P. W., Willoughby, R. H., O'Tuama, L. A., Swisher, C. N., and Benjamine, S. D.,** Neurological and intellectual sequelae of Reye's syndrome: a preliminary report, in *Reye's Syndrome I,* Pollack, J. D., Ed., Grune & Stratton, New York, 1975, 55.
11. **Bruner, L. R., O'Grady, D. J., Partin, J. C., and Schubert, W. K.,** Neuropsychologic consequences of Reye's syndrome, *J. Pediatr.,* 95, 706, 1979.
12. **Amla, I., Kamala, G. S., Gopalakrishma, G. S., Jayray, A. R., Sreenivasamurthy, V., and Parpia, H. A. B.,** Cirrhosis in children from peanut meal contaminated by aflatoxin, *Am. J. Clin. Nutr.,* 24, 609, 1971.
13. **Becroft, D. M. O. and Webster, D. R.,** Aflatoxins and Reye's disease, *Br. Med. J.,* 4, 117, 1972.
14. **Dalezios, J. I., Hsieh, D. P. H., and Wogan, G. N.,** Excretion and metabolism of orally administered aflatoxin B_1 by rhesus monkeys, *Food Cosmet. Toxicol.,* 11, 605, 1973.
15. **Lijinsky, W., Lee, K. Y., and Gallagher, C. H.,** Interaction of aflatoxin B_1 and G_1 with tissues of the rat, *Cancer Res.,* 30, 2280, 1970.
16. **Papageorgiou, A., Wigglesworth, F. W., Schiff, D., and Stern, L.,** Reye's syndrome in a newborn infant, *Can. Med. Assoc. J.,* 109, 717, 1973.
17. **Harris, H. B., Vogler, L. B., and Cassady, G.,** Reye's syndrome in a neonate, *South. Med. J.,* 69, 1511, 1976.
18. **Anderson, W. A. D.,** Zenker's degeneration, in *Pathology I,* Anderson, W. A. D., Ed., C. V. Mosby, St. Louis, 1971, 79.
19. **Peremans, J., DeGraef, P. J., Struble, G., and DeBlock, G.,** Familial metabolic disorder with fatty metamorphosis of the viscera, *J. Pediatr.,* 69, 1108, 1966.
20. **Satran, L., Sharp, H. L., Schenker, J. R., and Krivit, N.,** Fatal neonatal hepatic steatosis: a new familial disorder, *J. Pediatr.,* 75, 39, 1969.
21. **Butler, W. H. and Wigglesworth, J. S.,** Effects of aflatoxin B_1 on the pregnant rat, *Br. J. Exp. Pathol.,* 47, 242, 1966.
22. **Das, S. K., Nair, R. C., Patthey, H. L., and Mgbodile, M. U. K.,** The effects of aflatoxin B_1 on rat fetal lung lipids, *Biol. Neonate,* 33, 283, 1978.
23. **Pollack, J. D.,** Models of chemical and virus interaction and their relation to a multiple etiology of Reye's syndrome, in *Reye's Syndrome II,* Crocker, J. F. S., Ed., Grune & Stratton, New York, 1979, 341.
24. **Crocker, J. F. S., Rozee, K. R., Ozere, R. L., Digout, S. C., and Hutzinger, O.,** Insecticide and viral interaction as a cause of fatty visceral changes and encephalopathy in the mouse, *Lancet,* 2, 22, 1974.
25. **Colon, N. R., Ledesman, F., Pardo, V., and Sandberg, D. H.,** Viral potentiation of chemical toxins in the experimental syndrome of hypoglycemia, encephalopathy and visceral fatty degeneration, *Am. J. Digest. Dis.,* 19, 1091, 1974.
26. **Hug, G., Bosken, J., Bove, K., Linnemann, C. C., Jr., and McAdams, L.,** Reye's syndrome simulacra in liver of mice after treatment with chemical agents and encephalomyocarditis virus, *Lab. Invest.,* 45, 89, 1981.
27. **Svoboda, D., Reddy, J. K., and Chien L.,** Multinucleate giant cells in livers of marmosetts given aflatoxin B_1, *Arch. Pathol.,* 91, 452, 1971.
28. **Partin, J. C.,** Reye's syndrome (encephalopathy and fatty liver). Diagnosis and treatment, *Gastroenterology,* 69, 511, 1975.

29. Morgan, C., Howe, C., and Rose, H. M., Structure and development of viruses observed in the electron microscope, *J. Biophys. Biochem. Cytol.,* 2, 351, 1956.
30. Block, D. P., Morgan, C., and Godman, G. C., A correlated histochemical and electron microscopic study of the intranuclear ctystalline aggregates of adenovirus (RI-APC virus) in HeLa cells, *J. Biophys. Biochem. Cytol.,* 3, 1, 1957.
31. Crook, L. E., Harley, E. H., and Cohen, A., The mechanism of action of aflatoxin B_1: observation of virus-infected cells, *Chem.Biol. Interact.,* 5, 107, 1972.
32. Davis, L. E., Cole, L. L., Lockwood, S. J., and Kornfeld, M., Experimental influenza B virus toxicity in mice. A possible model for Reye's syndrome, *Lab. Invest.,* 48, 140, 1983.
33. Ladisch, S., Lovejoy, F. H., Hierholzer, J. C., Oxman, M. N., Strieder, D., Vawter, G. F., Finer, N., and Moore, M., Extrapulmonary manifestations of adenovirus type 7 pneumonia simulating Reye syndrome and the possible role of adenovirus toxin, *J. Pediatr.,* 945, 348, 1979.
34. Bradford, W. D. and Parker, J. C., Jr., Reye's syndrome. Possible causes and pathogenic pathways, *Clin. Pediatr.,* 10, 148, 1971.
35. Riley, H. D., Reye's syndrome, *J. Infect. Dis.,* 125, 77, 1972.
36. Reye, R. D. K., Morgan, G., and Baral, J., Encephalopathy and fatty degeneration of the viscera: a disease entity in childhood, *Lancet,* 2, 749, 1963.
37. Corey, L., Rubin, R. J., and Hattwick, M. A. W., A nationwide outbreak of Reye's syndrome: its epidemiologic relationship to influenza B, *Am. J. Med.,* 61, 615, 1976.
38. Linnemann, C. C., Jr., Shea, L., Partin, J. C., Schubert, W. K., and Schiff, G. M., Reye's syndrome: epidemiology and viral studies 1963—1974, *Am. J. Epidemiol.,* 101, 517, 1975.
39. Wogan, G N., Chemical nature and biological effects of the aflatoxins, *Bacteriol. Rev.,* 30, 460, 1966.
40. Bourgeois, C. H., Shank, R. C., Grossman, R. A., Johnson, D. O., Wooding, W. L., and Chandavimol, P., Acute aflatoxin B_1 toxicity in the macaque and its similarities to Reye's syndrome, *Lab. Invest.,* 24, 206, 1971.
41. Dvořáčková, I., Kusák, V., Veselý, D., Veselá, D., and Nesnídal, P., Aflatoxin and encephalopathy with fatty degeneration of viscera (Reye), *Ann. Nutr. Aliment.,* 31, 977, 1977.
42. Partin, J. C., Schubert, W. K., and Partin, J. S., Mitochondrial ultrastructure in Reye's syndrome, *N. Engl. J. Med.,* 285, 1339, 1971.
43. Cysewski, S. J., Pier, A. S., Baetz, A. L., and Cheville, N. F., Experimental equine aflatoxicosis, *Toxicol. Appl. Pharmacol.,* 65, 354, 1982.
44. Schwartz, A. D., The coagulation defect in Reye's syndrome, *J. Pediatr.,* 78, 326, 1971.
45. Doerr, J. A., Wyatt, R. D., and Hamilton, P. B., Impairment of coagulation function during aflatoxicosis in young chickens, *Toxicol. Appl. Pharmacol.,* 35, 437, 1976.
46. Clark, J. D., Jain, A. V., and Hatch, R. C., Effects of various treatments on induced chronic aflatoxicosis in rabbits, *Am. J. Vet. Res.,* 43, 106, 1982.
47. Kamden, L., Magdalou, J., and Siest, G., Effects of aflatoxin B_1 on the activity of drug-metabolizing enzymes in rat liver, *Toxicol. Appl. Pharmacol.,* 60, 570, 1981.
48. Thurlow, P. M., Desai, R. K., Newberne, P. M., and Brown, H., Aflatoxin B_1 acute effects on three hepatic urea cycle enzymes using semi–automated methods: a model for Reye's syndrome, *Toxicol. Appl. Pharmacol.,* 53, 293, 1980.
49. Tung, H. T., Donaldson, W. E., and Hamilton, P. B., Alterered lipid transport during aflatoxicosis, *Toxicol. Appl. Pharmacol.,* 22, 97, 1972.
50. Chaves-Carballo, E., Carter, G. A., and Wiebe, D. A., Triglyceride and cholesterol concentrations in whole serum and in serum lipoproteins in Reye syndrome, *Pediatrics,* 64, 592, 1979.
51. Miller, D. M., Crowell, W. A., and Stuart, B. P., Acute aflatoxicosis in swine: clinical pathology, histopathology, and electron microscopy, *Am. J. Vet. Res.,* 43, 273, 1982.
52. Mitchell, R. A., Ram, M. L., Arcinue, E. L., and Chang, C. H., Comparison of cytosolic and mitochondrial hepatic enzyme alterations in Reye syndrome, *Pediatr. Res.,* 14, 1216, 1980.
53. Faraj, B. A., Newman, S. L., Caplan, D. B., Ali, F. M., Camp, V. M., and Ahmann, P. A., Evidence for hypertyraminemia in Reye's syndrome, *Pediatrics,* 64, 76, 1979.
54. Lloyd, K. G., Davidson, L., Price, K., McLung, H. J., and Gall, D. G., Catecholamine and octopamine concentrations in brains of patients with Reye syndrome, *Neurology,* 27, 985, 1977.
55. Faraj, B. A., Caplan, B. D., Malveaux, E. J., Camp, V. M., and Ali, F. M., Similarity between tyramine-induced neurotixicity and the coma of Reye's syndrome, *J. Pharmacol. Exp. Ther.,* 226, 608, 1983.
56. Ikegwuonu, F. I., The neurotoxicity of aflatoxin B_1 in the rat, *Toxicology,* 28, 247, 1983.
57. Coulombe, R. A., Jr. and Sharma, R. P., Effect of repeated dietary exposure of aflatoxin B_1 on brain biogenic amines and metabolites in the rat, *Toxicol. Appl. Pharmacol.,* 80, 496, 1985.
58. Krishnamachari, K. A. V. R., Bhat, R. V., Nagarajan, V., and Tilak, T. B. G., Hepatitis due to aflatoxicosis, *Lancet,* 1, 1061, 1975.

59. **Ngindu, A., Kenya, P. R., Ocheng, D. M., Omondi, T. N., Ngare, W., Gatei, D., Johnson, B. K., Ngira, A. J., Nandwa, H., Jansen, A. J., Kaviti, N. J., and Siongok, T. A.,** Outbreak of acute hepatitis caused by aflatoxin poisoning in Kenya, *Lancet,* 1, 1346, 1982.
60. **Rozee, K. R., Lee, S. H. S., Crocker, J. F. S., Digout, S., and Arcinue, E.,** Is a compromised interferon response an etiologic factor in Reye's syndrome?, *Can. Med. Assoc. J.,* 126, 798, 1982.
61. **Miller, D. M.,** Beagle pup model of Reye's syndrome, *J. Natl. Reye's Syndrome Found.,* 6, 38, 1986.
62. **Mullen, P. W.,** Immunopharmacological considerations in Reye's syndrome: a possible xenobiotic initiated disorder?, *Biochem. Pharmacol.,* 27, 145, 1978.
63. **Millikin, P. D.,** Epithelial germinal centers: an acquired immunologic deficit?, *Am. J. Clin. Pathol.,* 81, 240, 1977.
64. **Dvořáčková, I., Vortel, V., and Hroch, M.,** Encephalitic syndrome with fatty degeneration of the viscera, *Arch. Pathol.,* 81, 240, 1966.
65. **Hanson, P. A. and Urizar, R. E.,** Ultrastructural lesions of muscle and immunofluorescent deposits in vessels in Reye's syndrome: a preliminary report of serial muscle biopsies, *Ann. Neurol.,* 1, 431, 1977.
66. **Tang, T. T., Harb, J. M., Grossberg, S. E., Sedmak, G. V., And Murphy, J. V.,** Leucocyte tubuloreticular inclusions in Reye's syndrome, *Arch. Pathol. Lab. Med.,* 109, 543, 1985.
67. **Rozee, K. R., Fernandez, R. C., Lee, S. H. S., Digout, S., and Crocker, J. F. S.,** The etiologic significance of a compromised interferon response in Reye's syndrome, in *Reye's Syndrome IV,* Pollack, J. D., Ed., Reye's Syndrome Foundation, Bryan, OH, 1985, 107.
68. **Thaxton, J. P., Tung, H. T., and Hamilton, P. B.,** Immunosuppression in chickens by aflatoxin, *Poult. Sci.,* 53, 721, 1974.
69. **Edds, G. T., Nair, K. P. C., and Simpson, C. F.,** Effect of aflatoxin B_1 on resistance in poultry against cecal coccidiosis and Marek's disease, *Am. J. Vet. Res.,* 34, 819, 1973.
70. **Thurston, J. R., Baetz, A. L., Cheville, N. F., and Richard, J. L.,** Acute aflatoxicosis in guinea pigs: sequential changes in serum proteins, complement, C_4 and liver enzymes and histopathologic changes, *Am. J. Vet. Res.,* 41, 1272, 1980.
71. **Chao-Fu, C. and Hamilton, P. B.,** Impaired phagocytosis by heterophils from chickens during aflatoxicosis, *Toxicol. Appl. Pharmacol.,* 48, 459, 1979.
72. **Hahon, N., Booth, J. A., and Stewart, J. D.,** Aflatoxin inhibition of viral interferon induction, *Antimicrob. Agents Chemother.,* 16, 277, 1979.
73. **Sullivan-Bolyai, J. Z. and Corey, L.,** Epidemiology of Reye syndrome, *Epidemiol. Rev.,* 3, 1, 1981.
74. **Hurwitz, E. S., Nelson, D. B., Davis, C., Morens, D. M., and Schonberger, L. B.,** National surveillance for Reye syndrome: a five year review, *Pediatrics,* 70, 895, 1982.

Chapter 4

CARCINOGENIC EFFECT OF AFLATOXIN ON MAN

I. INTRODUCTION

In the past 25 years, a number of very potent carcinogens of natural origin have been found, and the possibility that some of them may be involved in human cancer, particularly liver cancer, has led to extensive studies. This group of natural carcinogens included aflatoxin, which has been recognized as one of the most potent hepatocarcinogens in a wide variety of laboratory and farm animals such as ducks,[1] rainbow trout,[2] rats,[3] guppies,[4] rhesus monkeys,[5] and tree shrews.[6]

Furthermore, numerous investigations concerning the naturally occurring aflatoxin in foods and feeds have been reported throughout the world,[7-16] and a large-scale international effort has been made to determine what role aflatoxin plays in the etiology of human cancer.

II. PRIMARY LIVER CANCER IN THIRD WORLD COUNTRIES

The possibility that contamination of dietary staples by aflatoxin could be an etiological factor in liver cancer was originally suggested by LeBreton, Frayssinet, and Boy[17] in 1962. Later, Oettle[18] used intercountry comparisons to evaluate the possible association with a number of suspected etiological factors, such as kwashiorkor, malaria, hemosiderosis, and schistosomiasis, and finally concluded that a mycotoxin-contamination hypothesis fitted the known liver cancer data better than any other suspected factor. This hypothesis was formulated at a time when the carcinogenicity of aflatoxin was recognized, but there was only superficial circumstantial evidence as to its role in human cancer. Subsequent epidemiological data from various developing countries in Africa[10-16] and Southeast Asia,[7-9] from areas with a high incidence rate of liver cancer, revealed a definite correlation between the risk of developing liver cancer in man and the degree to which food products were contaminated with aflatoxin. The results of these studies were evaluated by the World Health Organization (WHO) in 1979.[19]

A. FACTORS IMPLICATED IN LIVER CARCINOGENESIS
1. Aflatoxin

The studies conducted by Keen and Martin[10,11] in Swaziland and Alpert et al.[12] in Uganda in 1971 were among the first epidemiological studies of the relationship between aflatoxin and liver cancer in Africa. In Swaziland, it had already been reported that the risk of developing liver cancer in humans varied with altitude. Stored groundnuts, a popular food in Swaziland, were often found to be contaminated with aflatoxin and the occurrence of such contamination was associated with an increased liver cancer frequency.[11]

Similar regional differences in the frequency of aflatoxin contamination of food have been reported in Uganda.[12] The detectable aflatoxin contamination of food samples (range: 10.8 to 43%) was associated with the increased incidence of liver cancer (range: 1.4 to 15.0) cases per 100,000 of the total population per year. As evaluation of the contamination of diets by aflatoxin and the liver cancer incidence in the African population of the Murang'a district of Kenya established a statistically significant correlation.[13] Further studies of a similar design, carried out in Mozambique,[16] Swaziland,[14] and Zaire,[20] have proved a clear correlation between actual concentrations of aflatoxin in meals and the liver cancer incidence.

Aflatoxin has been found contaminating a number of foodstuffs in Thailand.[7-9] The most frequently contaminated food was peanuts and products derived from them. The attempt to correlate the average aflatoxin consumption with the incidence of liver cancer in Thailand was initiated by a team conducted by Shank et al.[7-9] in 1967. The studies revealed a close correlation between the liver cancer incidence and the amount of aflatoxin consumed by the inhabitants.

Later studies were carried out by several groups of investigators in Thailand and were reviewed in 1981.[21] To verify the causal correlation between aflatoxin consumption and the liver cancer incidence, extensive studies were performed in selected villages of three provinces of Thailand: Singburi, Ratburi, and Songkhla. An aflatoxin survey was done in 1967 to 1969, the liver cancer incidence was recorded between 1978 to 1981. A review of the collected data from these studies revealed that the highest incidence of liver cancer occurred in Singburi with an incidence rate of $5.8/10^5$ of the population per year, which correlated with the highest average of the total of aflatoxin ingested (73 to 81 ng/kg/d), in contrast to Ratburi with a cancer incidence rate of $3.7/10^5$ of population and an average total aflatoxin intake of 45 to 77 ng/kg/d. The lowest incidence of liver cancer was found in Songhkla, with an incidence rate of $1.7/10^5$ of population per year, where concurrently the lowest aflatoxin daily intake (5 to 8 ng/kg/d) was found.

Carlborg[22] in his mathematical model based on an epidemiological data analysis of studies in Africa and Southeast Asia obtained a linear dose-response curve between the lifetime risk of liver cancer and the daily aflatoxin intake in male population. The lifetime risk was expressed as the number of deaths per 100,000 due to liver cancer in a life span of 54 years in male population. The daily aflatoxin intake was in ng/kg of body weight per day. From this analysis, a background risk at zero aflatoxin intake was determined to be 121 deaths per 100,000 and an increment of risk due to ingestion of 1 ng/kg/d was determined to be about 10 deaths per 100,000.

Hsieh and Ruebner,[23] however, have argued that the cancer profiles vary widely between countries and the values reported by Carlborg[22] do not hold true for all human population. Using the U.S. male population as an example, they have reported that an annual death rate due to liver cancer is 3.24/100,000, and the lifetime risk due to liver cancer represents 227 deaths per 100,000, on the assumption that the life span of U.S. males is 70 years. Based on epidemiological data, the level of the lifetime risk experienced by U.S. males corresponds to an exposure level of 10.6 ng/kg/d. However, the level of aflatoxin exposure in the U.S. population, as calculated from average AFB concentrations in corn and peanut products and the consumption of these foods, is 20.0 ng/kg/d, a level about twice as high as the expected value. The difference between the expected and the estimated exposure levels suggests that U.S. males have to be twice as resistant to AFB_1 hepatocarcinogenicity as are males in Africa and Asia. Therefore, they assumed that differences in the relative significance of liver cancer in the population of other countries than those of Africa and Asia cannot be totally accounted for by differences in the levels of ingested aflatoxin, and other factors must be considered in the etiology of liver cancer.

2. HBV Infection

Epidemiological studies have strongly implicated chronic hepatitis B virus infection as a further risk factor in the induction of primary liver cancer.[24,25] In some areas of Africa and Southeast Asia, the carrier rate of HBV — even in children — has been estimated to be higher than 30%.[26] A prospective study of 22,707 men in Taiwan confirmed that a chronic HBV infection precedes tumor development and increases the incidence of liver cancer to 1150 cases per 100,000 men at risk, compared with 5 cases per 100,000 uninfected controls.[27] In Uganda and Zambia, 96% of patients with liver cancer have been reported to have markers of HBV infection.[28] A strong association between liver cancer and HBV infection has also been noted in intertropical Africa, in Senegal.[29] Out of 103 patients with liver cancer, 80 had an active HBV infection and 23 showed signs of a previous HBV infection. The HBV-PLC relationship in Senegal seems to be well-established in patients less than 50 years old. However, the frequency of HBV replication markers in patients more than 50 years old was no different from the frequency observed in leprosy patients and blood donors of the same region and age group. Therefore, it has been assumed that HBV infection, which often appears early in the life of people in Senegal, was followed by a long-lasting HBV replication that led to the development of liver cancer, while in the older patients with HBV serum markers of a previous infection, the initial HBV replication induced liver damage which might progress to liver cancer occurring late in life, or that other agents might be responsible for cancer development in older patients.

Recently FuSun Yeh et al.[30] reported that 80% of liver cancer cases in Guangxi (autonomous region of China) were HBsAg positive compared to 22% of the controls. They have concluded that HBV infection can account for at least 80% of all liver cancer cases occurring in Guangxi. Their findings coincide with those reported from Hong Kong[31] and Taiwan.[32]

However, another recently conducted study in China[33] demonstrated marked geographical differences in the liver cancer incidence. The areas with the highest incidence of the tumor were largely rural and serological surveys conducted in high and low liver-cancer incidence areas have failed to demonstrate proportional differences in the prevalence of the HBV carrier rate between these regions. However, marked differences in the level of dietary exposure to aflatoxin have been noted, the rural areas having higher rates of exposure to aflatoxin.

Even though many arguments favor HBV as an etiological agent of PLC, at present there is no clear evidence that HBV as such has a neoplastic transforming activity in cultured cells or experimental animals,[34,35] in contrast to aflatoxin which has been shown to increase the incidence of liver cancer in at least eight species of experimental animals.[36] In addition, it has been shown *in vitro* experiments that the metabolism of aflatoxin with the microsomal fraction of human hepatocytes is identical to the aflatoxin metabolism with a microsomal fraction of the rat liver, and in particular transformation to 2,3-oxide, which has been implicated as the ultimate carcinogenic metabolite of AFB_1.[37]

3. Synergistic Mechanism Between Aflatoxin and HBV

To date it is being increasingly accepted that the ingestion of aflatoxin and a chronic HBV infection — both widespread in the Third World countries — are implicated in the etiology of primary liver cancer.[38] The nature of the synergistic mechanism between these agents is not yet clear and remained an object of hypotheses. It is known that aflatoxin is metabolized within the endoplasmic reticulum by the microsomal mixed-function oxidase (MFO) system leading to the production of reactive metabolites of AFB_1. Some of these metabolites may be detoxified to biologically inactive compounds, whereas others may bind covalently with liver cell DNA.[39] It has been suggested[40] that replication of the altered genetic material, resulting from the irreversible covalent binding of AFB_1-2,3-epoxide to liver cell DNA, may induce a distinct predisposition to cancer development and ultimately lead to liver cancer. A subsequent integration of HBV DNA in such previously damaged cells may potentiate tumor development. The enhanced severity of liver lesions reported in animals injured with viral hepatitis, superimposed upon aflatoxin injury, supports this hypothesis.[34,35] A long-term observation[34] of the effect of AFB_1 and viral hepatitis on the marmoset liver revealed that (1) viral hepatitis alone did not induce liver cirrhosis and/or hepatic tumors; (2) in the case of AFB_1 injury, the marmosets had cirrhosis in addition to hepatic tumors; and (3) double injury (AFB_1 and viral hepatitis) produced more severe effects on the liver than when the animals were injured with a single agent.

Alternatively, prior damage to the liver due to viral hepatitis renders the organ more susceptible to a hepatotoxic agent.[41] Human hepatocytes infected with HBV show hypertrophy of the smooth endoplasmic reticulum, an ultrastructural change which has now come to be regarded as a morphological expression of the drug-induced enzyme production in the liver.[42] Thus such virus-modified cells may potentiate the transformation of aflatoxin to its highly reactive metabolite or suppress the production of nonreactive metabolites.[41]

The hypothesis proposed by Lutwick[38] suggests that aflatoxin may act as an immunosuppressive agent rather than a primary carcinogen, as it seems to do in animals. AFB_1 is known to reduce resistance to various bacterial and viral infections,[43-45] and alters the immune system in many ways: it affects antibody formation[46] and phagocytosis,[47] reduces the migration inhibition factor and phytohemagglutin stimulation of lymphocytes in aflatoxin-treated guinea pigs,[48] depresses the phytohemagglutin stimulation of human lymphocytes *in vitro*,[49] and inhibits viral interferon induction.[50]

These alterations of the immune system may increase the HBV carrier rate, and hence the risk of liver cancer; alternatively, an aflatoxin-impaired immune surveillance system might not be able to detect and eradicate malignant foci.[38]

4. Additional Risk Factor

At present, epidemiological studies have mainly focused on aflatoxin and HBV infection in liver cancer induction. However, additional risk factors of etiological importance must be considered. The susceptibility of a population to liver cancer is also known to be influenced by its lifestyle, nutritional conditions, and exposure to other xenobiotic factors.[51] A recent case-control investigation from the Philippines[52] has emphasized the potential role of alcohol in an interactive role involving a persistent HBV infection and exposure to aflatoxin in liver cancer genesis.

A possible synergistic role played by AFB_1 and nitroamines in liver carcinogenesis has been suggested.[53] Investigations for amines in the environment have proven the presence of nitrates in a wide variety of foodstuffs,[54] and their enhancing effect on aflatoxin-induced liver cancer in animals has recently been reported.[55]

III. PRIMARY LIVER CANCER IN THE WESTERN COUNTRIES

Not long ago, primary liver cancer was regarded as relatively uncommon in Europe and North America in contrast to Africa and Asia, where this tumor may constitute 30 to 50% of all carcinomas.[56,57] Numerous reports, however, have demonstrated a gradual increase in the incidence of this tumor in the Western world during the past 20 years.[58-63] MacSween[59] in his review of primary malignant liver tumor over a period of 70 years in Glasgow found a steady increase in their incidence since the 1940s, occurring from 0.27% (1940 to 1949) to 0.52% in 1960 to 1970. A similar twofold increase in the liver-tumor incidence was later reported by Burnett.[64] The cause is not clear and no convincing explanation has been offered for the increase in liver cancer incidence in the Western countries till now.

A. FACTORS IMPLICATED IN LIVER CARCINOGENESIS
1. Cirrhosis

In some of the studies it has been suggested that an increase in liver cancer cases has followed an increase in the incidence of cirrhosis.[59] Other researchers, however, argued that an increase in tumor incidence appeared to be independent of cirrhosis incidence, which remained approximately stable.[58,64] Glenert[60] and Ohlsson and Norden,[61] in Scandinavia and Elkington et al.[62] in the U.K. reported a significant increase in liver tumor incidence, but were unable to find evidence that a simultaneous cirrhosis was of importance. Patton and Horn[63] postulated that an increase in liver cancer may be due to longer survival of cirrhotic patients because of improved treatment. Burnett et al.,[64] however, in their 25-year-long review of liver cancer in Scotland, found no differences between the mean age at death in either cirrhotic or liver cancer patients.

The association of liver cancer with cirrhosis has long been recognized, but the nature of this relationship remains a matter of some speculation. Gall[65] drew attention to the particular predilection liver cancer has for the macronodular type of cirrhosis, believed to be post-hepatitic and post-necrotic. This observation has been confirmed by several subsequent studies[57-59,63] and has led to the suggestion that the development of cancer reflects a greater propensity to hyperplasia, which appears to be a feature of macronodular cirrhosis. It has been assumed[64] that the continuing parenchymal destruction, implicit in the cirrhotic process, is capable of provoking a hepatocyte regeneration and hyperplasia to the start of a new growth of autonomous character. This hypothesis is supported by the observation that liver cancer developing in cirrhosis is often associated with morphological evidence of an active hepatitis, which indicates a long-standing and continuing regeneration and hyperplasia of the hepatocytes.[59,64] Although the frequent association of PLC with cirrhosis is widely accepted, it cannot explain, for example, the occurrence of tumors in noncirrhotic livers and other factors must be considered to be involved in tumor genesis. It has been admitted that cirrhosis may act as an important promoting agent, but most probably in the presence of a primary carcinogenic factor (or factors), either viral or chemical.[64]

2. Alcohol

As regards the causal factor, chronic alcoholism has been considered the most commonly known etiological agent in the Western countries in the genesis of both cirrhosis and liver cancer. Up to one third of the alcoholic patients developed PLC,[66] especially those who had given up alcohol before death and developed a coarse granular type of cirrhosis. This view, has not been however, confirmed by others. In a series of 520 cirrhotic patients reported by MacSween and Scott,[67] 96 of them had a history of alcohol abuse, but only 9 of them had liver tumors. In his subsequent study of 100 cases of liver cancer, a fatty or nutritional cirrhosis, which was most commonly associated with alcoholism, was not found in any of the examined cases.

3. HBV Infection

Recent studies of liver cancer in the European or American population have demonstrated a striking association of HBV infection with PLC.[68-70] Yarrish et al.[70] recently found a chronic HBV infection in 50% of liver cancer patients in Chicago. A past or present HBV infection in liver cancer patients in Great Britain has been found common, although the frequency of HBV in the general British population is low.[69] Even in Greece,[68] the prevalence of coincident HBsAg and anti-HBs (with or without anti HBc) was higher in the liver cancer patients than in the controls. In a series of 80 cases of liver cancer in Caucasian patients in Greece, 48% of them had markers of HBV infection. Active infection has been found more common in PLC patients with coexisting cirrhosis than in those without cirrhosis (67 vs. 26%).

Although the widely held view is that association of HBV with PLC is much stronger among patients with cirrhosis, this has not always been noted. Lutwick[71] reported that 25% of HBV – associated hepatoma do not demonstrate either cirrhosis or fibrosis histologically.

The frequent association of HBV infection with liver cancer, found even in the Western countries where the incidence of HBV infection in general population is low, led to the view that the HBV represents an oncogenic factor etiologically related to liver cancer. It has been postulated that the association of the HB virus with liver cancer can be explained in one of the three ways:[68] (1) chronic HBV infection, in association with other carcinogens, could result in PLC; (2) PLC could facilitate persistent HBs-antigenemia when the individual is exposed to HBV; (3) other toxic, infectious, genetic, or nutritional factors may affect both the genesis of PLC and the increased response to HBV.

IV. CASE CONTROL STUDY OF LIVER CANCER IN CZECHOSLOVAKIA

A significant increase in the incidence of PLC has been noted in the necropsy material of the Faculty Hospital in Hradec Králové, Czechoslovakia, during the past 20 years. This fact stimulated a systematic study of this tumor which was prior seen very rarely in the autopsy material of the institute. The study was directed to the frequency of liver cancer and its histological type, tumor incidence in a cirrhotic and noncirrhotic liver, evidence of hepatitis infection, and alcohol consumption. In addition, investigation for the presence of aflatoxin was carried out to evaluate its possible risk of cancer development under conditions different from those seen in developing countries.

A. MATERIALS AND METHODS

Clinical and necropsy records of 181 PLC patients dying in this hospital between 1951 and 1985 were examined. The histopathology of every tumor was reviewed to determine its type. A note was made of the type of cirrhosis associated with malignancy. Figures relating to the incidence of liver cirrhosis without tumor during the period under investigation were also reviewed.

1. HBsAg Determination

Histological sections of cancerous and noncancerous liver tissue were stained with Orcein according to Shikata et al.[72] and Victoria blue according to the method described by Tanaka et al.[73] Sera of the patients have been systematically investigated for HBsAg since 1970 by reverse passive hemagglutination (RPHA) and enzyme-linked immunosorbent assay (ELISA) tests. Other serological markers of hepatitis B virus infection could not be performed because the reagents were not available.

2. Aflatoxin Determination

Liver samples of both cancerous and noncancerous tissue taken at the autopsy of 72 cases were assayed by chromatographic technique (TLC) and by radioimmunoassay (RIA). The procedures have been described in detail previously. Liver samples of 55 patients, 30 men and 25 women aged 50 to 70 years, who died of unrelated causes (10 of liver cirrhosis, 15 of an acute heart failure, 10 of gastric and 10 of colon carcinomas, 5 of cerebral hemorrhages, and 5 in car accidents) served as the controls.

Liver samples of eight patients with liver cancer (seven hepatocellular, one cholangiocarcinoma) were supplied to Dr. Garner, (the Cancer Research Unit, University of York, U.K.) for ELISA.[74]

Liver-sample preparation — Liver samples were homogenized in 1.5 nM sodium citrate (15 nM NaCl), precipitated with 3 volumes of ethanol, centrifuged, and the pellets freeze-dried for shipment to York. Samples were then stored at –90°C until analyzed. For analysis, the dried liver sample was resuspended in 1 nM EDTA (1%) SDS and left stirring on ice until completely dissolved. Thereafter, DNA isolation was performed using the procedure of Stanton et al.[75] Samples were hydrolyzed at 90°C in 5% trichloroacetic acid; an aliquot taken for liquid scintillation counting and DNA concentration was estimated using the Burton diphenylamine procedure.

ELISA — The samples were analyzed by ELISA using a monoclonal antibody (6 E9) isolated by Microtest Research Ltd., York, U.K. The assay procedure, as previously reported, used a second anti-mouse antibody conjugated to peroxidase,[76] or a commercially available 96-w ll Kit for aflatoxin analysis, using an enzyme amplification system (Aflatoxin Immunoassay Kit, Microtest Research Ltd., York, U.K.).

B. RESULTS AND COMMENTS

Tumor incidence and relating data of patients with PLC are presented in Figures 64 and 65. The results obtained on the basis of a 35-year-study of PLC in the hospital population in Hradec Králové were, in most instances, comparable to those reported by others.[58,59,63,64] Among the 34,110 autopsies, 181 (0.53%) cases of PLC have been found. The notable increase in tumor incidence from 0.23% (1951 to 1960) to 1.20% (1981 to 1985) was apparent. By contrast, the incidence of liver cirrhosis (approximately 1.33% per year) remained stable during the period under investigation. In 107 cases (59%) the tumors were associated with liver cirrhosis, largely

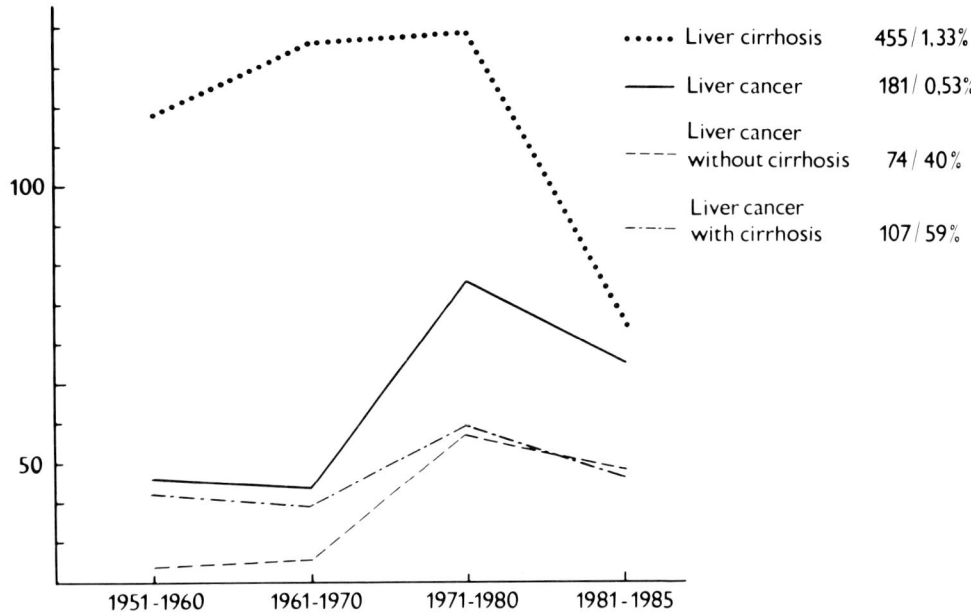

FIGURE 64. Incidence of primary liver cancer and cirrhosis, 1951 to 1985.

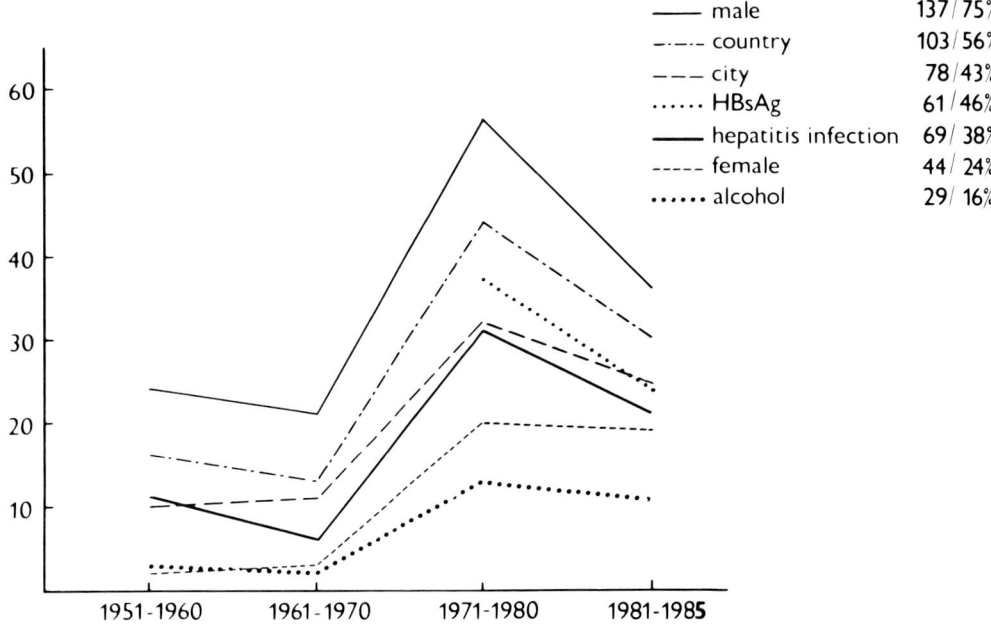

FIGURE 65. Data related to the patients with PLC, 1951 to 1985.

of the macronodular type (94%), while only in 8 cases the tumor was found in micronodular cirrhosis. An increased incidence of tumors in noncirrhotic livers has been noticed since 1970 (48%) by contrast with the previous years 1951 to 1970 (15 to 20%).

Liver cancer occurred more frequently in men than in women (75% vs. 24%), although an increase in liver cancer was observed in women over the last 20 years (from 7.6% in 1951 to 1960 to 34% in 1981 to 1985).

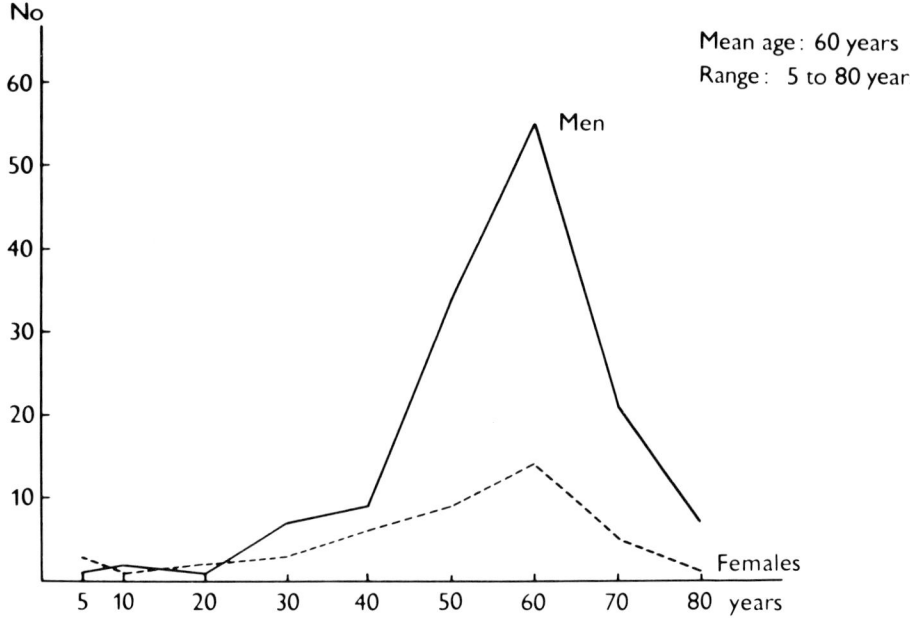

FIGURE 66. Age and sex distribution of patients with PLC, 1951 to 1985.

The mean age of the patients was between 50 and 60 years, ranging from 5 to 80 years (Figure 66). A slight prevalence of liver cancer patients has been observed in patients living in the country (56%) when compared to those living in the cities (43%).

1. Histological Investigation

The most frequent form of tumor found in 97 (53%) cases was a well-differentiated trabecular carcinoma. The tumor was composed of plates of polygonal cells arranged in cords varying from a few cells in thickness to as many as two dozen, circumscribed by endothelial cells. The plasm of certain cells contained fine granules of bile pigment (Figure 67). In some areas multinucleated cells were present (Figure 68). In 67 (37%) cases the tumor revealed trabecular structures which differentiated in some areas to acinous forms similar to thyroid-gland follicles. The acini were irregular in shape, variable in size, and contained eosinophilic proteinaceous substance (Figure 69). Multinucleated giant cells containing bizarre nuclei were frequently seen (Figure 70). Nine (4%) tumors were formed by thick columns, appearing as sheets consisting of irregularly arranged, usually undifferentiated epithelial cells. Areas of anaplastic spindle-shaped cells resembling sarcoma cells were present (Figure 71). Pure cholangiocarcinomas were found in 13 (7%) cases of this series. The tumor was formed by glandular or alveolar structures composed of cuboidal or columnar epithelial cells (Figure 72). The cytoplasm was of fine granular structure. The mucous substance stained positive with PAS, and Halle-Muller was present in the lumina of the glandular structures. In some areas the gland-like structures were enclosed in an abundant fibrous stroma resembling scirrhous formation (Figure 73).

The histological sections of the liver stained with Orcein and Victoria blue failed to detect HBsAg in both the tumor and the noncancerous liver tissue, even in cases where HBsAg was found serologically.

2. HBV Infection

Sixty-nine (38%) patients had a history of a previous hepatitis infection 5 to 15 years before the onset of a tumor. HBs-antigenemia was found in 59 (46%) of the 131 patients' sera

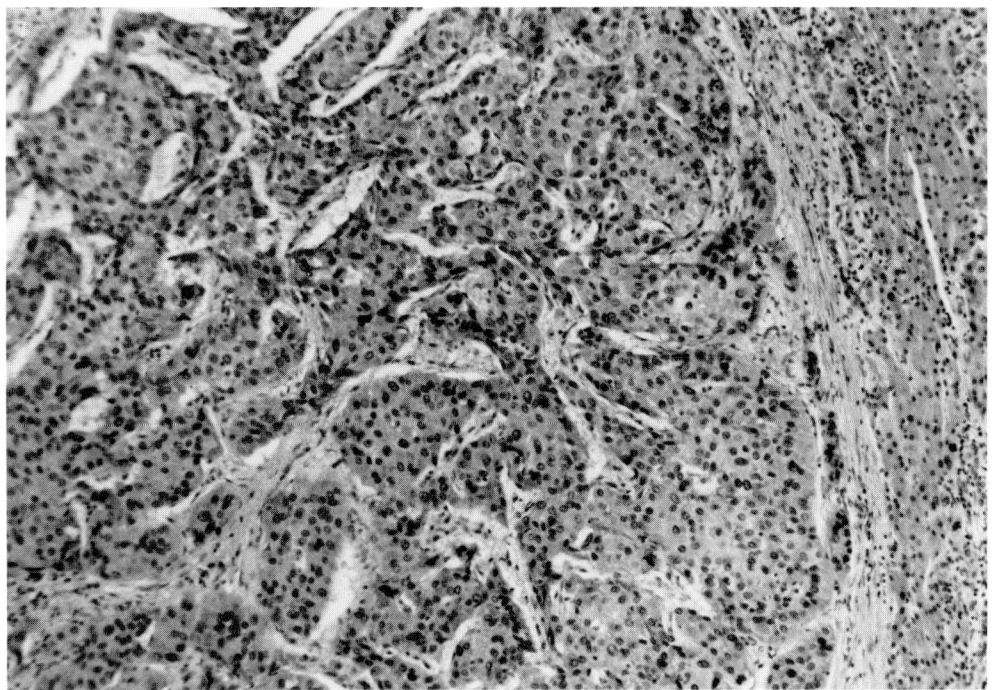

FIGURE 67. Well-differentiated trabecular hepatocellular carcinoma with thin endothelial lined clefts between trabeculae (hematoxylin and eosin, ×90).

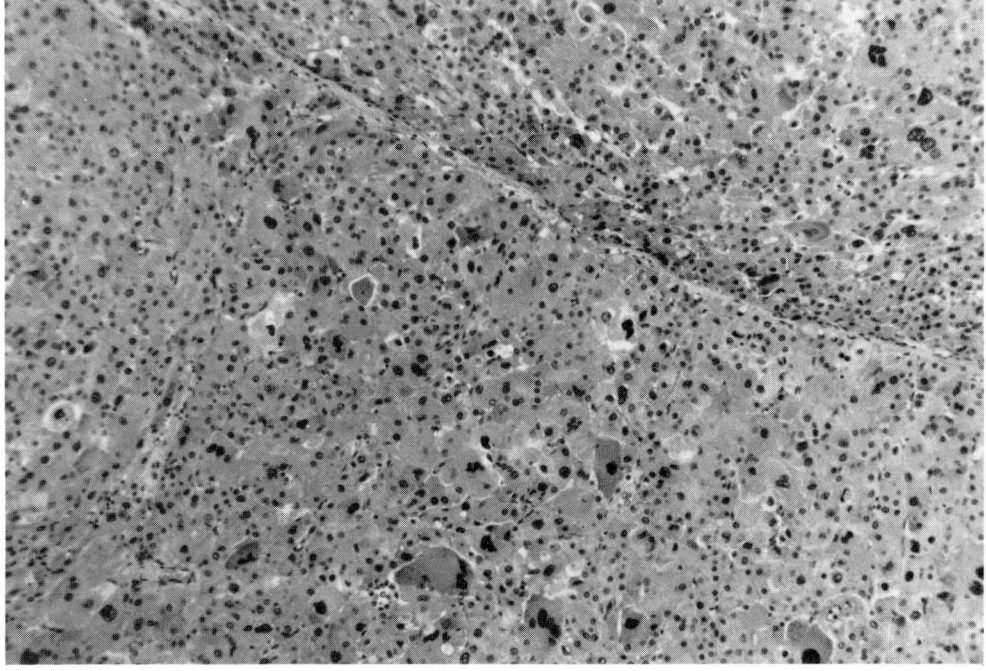

FIGURE 68. Moderately differentiated area of hepatocellular carcinoma showing the presence of giant cells (hematoxylin and eosin, ×140).

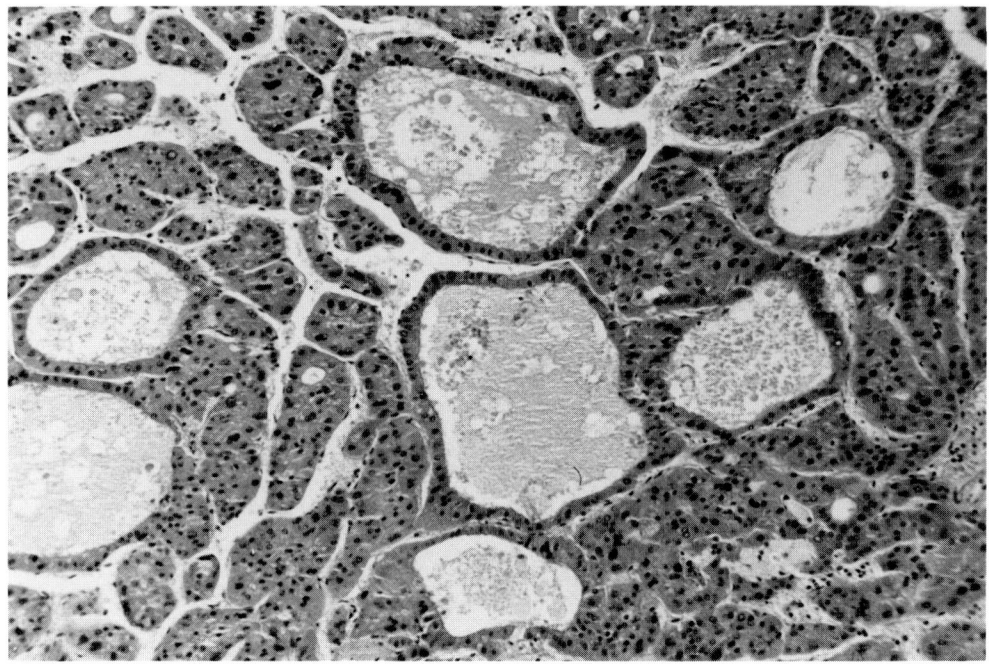

FIGURE 69. Hepatocellular carcinoma showing both trabecular and acinar pattern. The acini contain proteinaceous substance (hematoxylin and eosin, ×190).

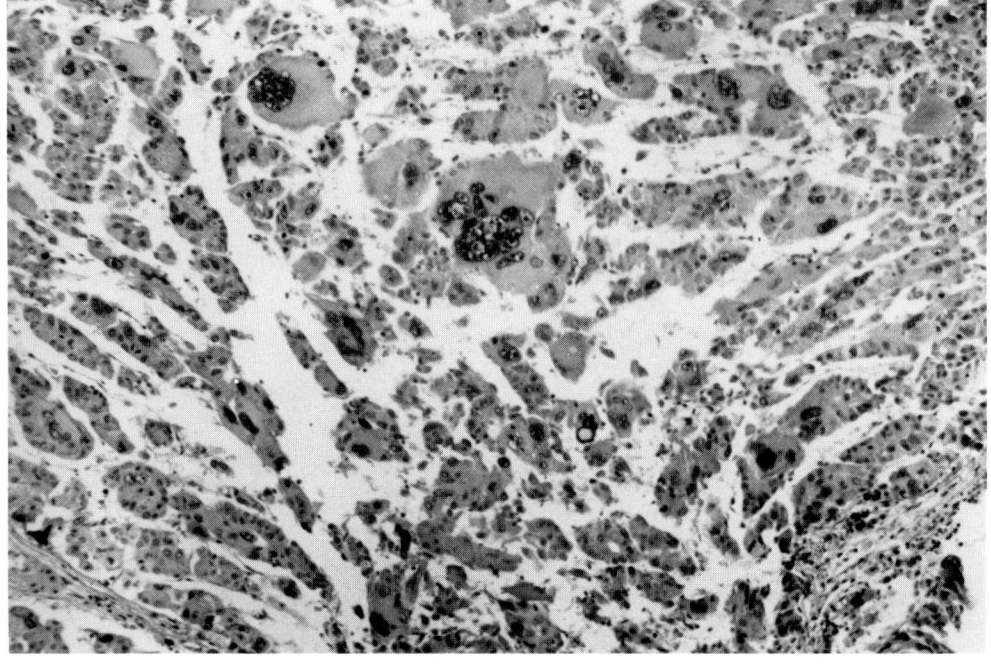

FIGURE 70. Hepatocellular carcinoma composed of columns poorly differentiated epithelial and giant multinucleated cells (hematoxylin and eosin, ×230).

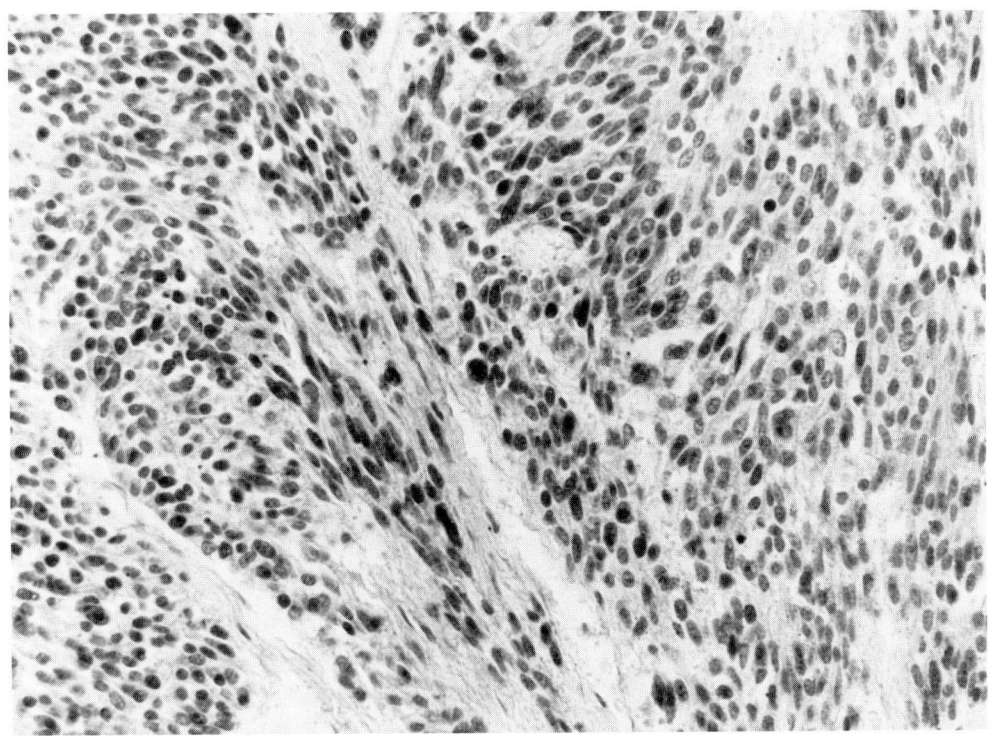

FIGURE 71. Spindle cell areas resembling sarcoma cells (hematoxylin and eosin, ×135).

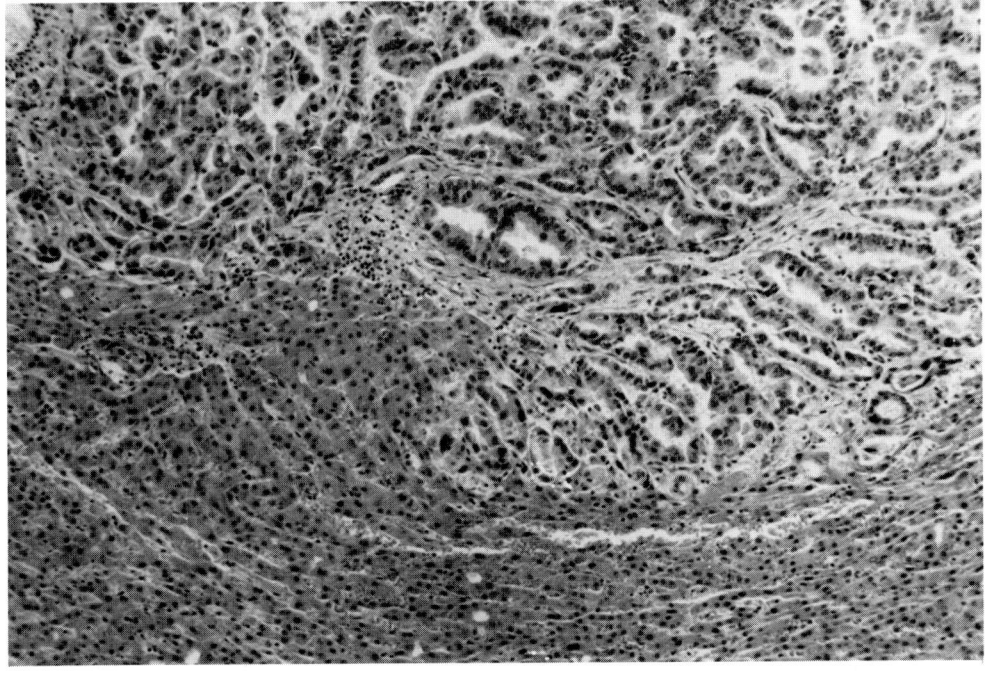

FIGURE 72. Cholangiocarcinoma with a few papillary infoldings (hematoxylin and eosin, ×100).

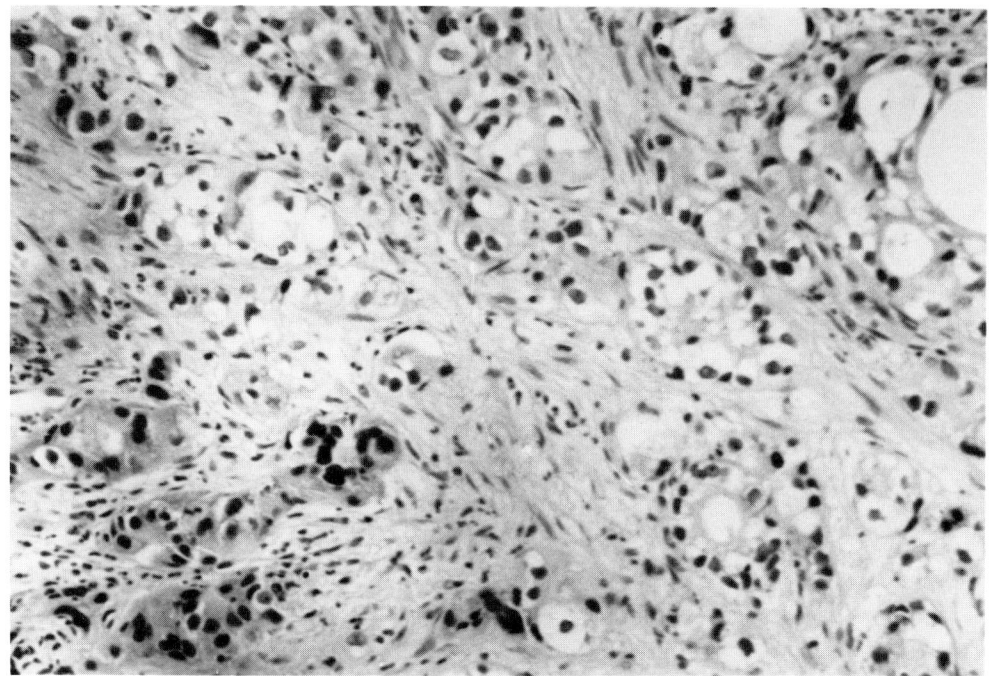

FIGURE 73. Cholangiocarcinoma with glandular structures enclosed in an abundant fibrous stroma (hematoxylin and eosin, ×140).

investigated for HBsAg. A significant association of liver cancer with HBV infection has been well-documented by a large studies in Africa and Asia,[77-81] where the incidence of HBV infection in general population is high. However, a past or present HBV infection, commonly associated with liver cancer, has recently been reported in the Western countries where the frequency of HBV in general population is low.[68,69] The number of liver cancer patients with HBsAg positivity found in this series was consistent with that reported in Greek liver cancer patients[68] and may indicate the suspected link between HBV infection and liver cancer.

3. Alcohol Consumption

Twenty-nine (16%) of the patients had a history of a chronic alcohol abuse, but only 13 of them had documented cirrhosis. None of them showed either a fatty cirrhosis or Mallory hyaline bodies, changes most commonly associated with alcoholism. Although alcohol consumption has been suggested as an important factor in liver cancer genesis in the Western countries,[68] the results of this study have not proved alcohol to be an important etiological agent.

4. Aflatoxin in Liver Cancer Patients

In 5 of 17 liver samples of PLC patients autopsied in the years 1972 to 1980, aflatoxin B_1 was found in an amount ranging from 1.3 to 5.2 ng/g when assayed by TLC and RIA (Table 34). Since 1981 liver samples of each PLC case taken at the autopsy were investigated for aflatoxin. AFB_1 was detected in 32 (58%) out of 55 liver samples at concentrations of 3.1 to 148 ng/g of the liver tissue. Tables 35 to 39 indicate the levels of AFB_1 in individual cases.

Aflatoxin, however, was not found in any of the 13 cholangiocarcinomas of this series, and HBsAg was present only in two of the patients bearing this tumor. This fact is not surprising. It is known that bile duct carcinoma arises from intrahepatic ducts, and its clinical and morphological features differ distinctly from those of primary hepatocellular carcinoma. Intrahepatic bile duct carcinoma is rarely associated with liver cirrhosis, shows no apparent

TABLE 34
Level of Aflatoxin in Liver Samples and Other Data Related to Five Patients with PLC (1972—1980)

No.	Sex	Age (years)	Domicile City	Domicile Country	IH	HBsAg	Alcohol	Ci	Type of tumor	AFB$_1$ (ng/g)
1	F	58		+					HC	5.1
2	F	80	+					+	HC	5.2
3	M	56	+		+	+		+	HC	5.0
4	M	54		+			+	+	HC	1.3
5	M	53	+		+	+			HC	1.6

Note: HC = hepatocellular carcinoma.

TABLE 35
Level of Aflatoxin in Liver Samples and Other Data Related to the Patients with PLC in 1981

No.	Sex	Age (years)	Domicile City	Domicile Country	IH	HBsAg	Alcohol	Ci	Type of tumor	AFB$_1$ (ng/g)
1	M	37		+	+	+		+	HC	Negative
2	F	56		+	+			+	HC	26.7
3	M	46	+		+	+			HC	17.6
4	M	50		+			+	+	HC	Negative
5	F	52		+		+			HC	11.6
6	M	58	+			+	+	+	HC	19.8
7	M	14		+	+				HC	Negative
8	F	69	+			+			HC	29.1
9	F	49	+						ChC	Negative
10	M	67	+		+	+		+	HC	Negative

Note: HC = hepatocellular carcinoma.
ChC = cholangiocellular carcinoma.

TABLE 36
Level of Aflatoxin in Liver Samples and Other Data Related to the Patients with PLC in 1982

No.	Sex	Age (years)	Domicile City	Domicile Country	IH	HBsAg	Alcohol	Ci	Type of tumor	AFB$_1$ (ng/g)
1	F	62	+			+			HC	Negative
2	M	63		+	+		+	+	HC	40.2
3	F	62		+					HC	9.7
4	M	57	+		+	+		+	HC	38.2
5	F	48	+						ChC	Negative
6	M	45		+	+		+	+	HC	37.2
7	F	68		+		+			ChC	Negative
8	M	78		+					HC	Negative
9	F	62		+		+			HC	23.6
10	M	72		+	+			+	HC	19.9

Note: HC = hepatocellular carcinoma.
ChC = cholangiocellular carcinoma.

TABLE 37
Level of Aflatoxin in Liver Samples and Other Data Related to the Patients with PLC in 1983

No.	Sex	Age (years)	Domicile City	Domicile Country	IH	HBsAg	Alcohol	Ci	Type of tumor	AFB$_1$ (ng/g)
1	M	58	+		+	+	+	+	HC	39.7
2	M	68	+			+		+	HC	27.3
3	M	82	+					+	HC	148.4
4	M	73		+				+	HC	37.6
5	M	30	+		+	+			HC	Negative
6	F	11		+				+	HC	17.6
7	M	45	+			+	+		HC	25.3
8	M	57	+		+	+		+	HC	Negative
9	F	5		+					HC	5.10
10	M	87	+		+			+	HC	29.1
11	F	88		+		+			HC	Negative
12	F	63		+					HC	Negative

Note: HC = hepatocellular carcinoma.

TABLE 38
Level of Aflatoxin in Liver Samples and Other Data Related to the Patients with PLC in 1984

No.	Sex	Age (years)	Domicile City	Domicile Country	IH	HBsAg	Alcohol	Ci	Type of tumor	AFB$_1$ (ng/g)
1	F	65	+						ChC	Negative
2	M	65		+	+			+	HC	5.42
3	M	64		+					HC	3.6
4	M	60	+		+		+	+	HC	9.0
5	M	69		+		+	+	+	HC	3.60
6	M	62		+		+		+	HC	3.50
7	M	62	+		+	+		+	HC	3.30
8	F	74		+					HC	Negative
9	M	86	+				+		HC	3.10
10	M	76		+					HC	Negative
11	F	45		+					HC	Negative

Note: HC = hepatocellular carcinoma.
ChC = cholangiocellular carcinoma.

difference in sex, and is rather rare in the tropics. Factors other than those implicated in the genesis of hepatocellular carcinoma, such as intrahepatic stones and parasitic infestation, are suggested to be responsible for the development of this tumor.[82]

Except for two patients with liver cirrhosis, in whom aflatoxin B$_1$ was found at levels of 1.2 and 2.3 ng/g, no aflatoxin was detected in the control liver samples.

The presence of AFB$_1$ found in the liver samples of PLC patients is of particular interest and provides evidence that the health risk of human exposure to aflatoxin is apparently not limited to the developing countries, as has been earlier suggested. Aflatoxin contamination of foodstuffs has not been generally considered a health hazard in advanced societies, where modern food processing and storage methods reduce the risk of ingestion of this toxin. However, this

TABLE 39
Level of Aflatoxin in Liver Samples and Other Data Related to the Patients with PLC in 1985

No.	Sex	Age (years)	Domicile City	Domicile Country	IH	HBsAg	Alcohol	Ci	Type of tumor	AFB$_1$ (ng/g)
1	M	32		+	+		+		HC	13.8
2	F	76		+					HC	17.4
3	M	61	+		+	+		+	HC	10.8
4	M	57		+		+		+	HC	Negative
5	M	42	+		+				HC	Negative
6	M	58	+				+		HC	Negative
7	M	60		+	+	+		+	HC	12.1
8	F	65		+				+	HC	10.9
9	M	66		+	+				HC	Negative
10	F	60	+						ChC	Negative
11	M	11	+					+	HC	5.3
12	M	73					+		HC	Negative

Note: HC = hepatocellular carcinoma.
ChC = cholangiocellular carcinoma.

assumption has been recently challenged. Cases bearing evidence of aflatoxin in the liver of patients with liver cancer were occasionally reported in the U.S.[83,84] Furthermore, the U.S. Food and Drug Administration has estimated that aflatoxin is responsible for 1 case of cancer per 100,000 of population in the U.S.; this assumes an aflatoxin load of 2 ppb in peanuts and of 10 ppb in corn products.[85]

In all liver samples investigated by TLC and RIA, only AFB$_1$ — none of its metabolites — was detected. In animal experiments, however, it was shown that orally administered AFB$_1$ undergoes an early metabolism, and unmetabolized AFB$_1$ is excreted during the first day after exposure. The presence of AFB$_1$ in the human liver may thus indicate either a recent toxin exposure or impaired metabolizing activity due to a previously damaged liver. The latter hypothesis seems to be more plausible. It has been demonstrated that during a protein deficiency in malnourished children, a large proportion of ingested AFB$_1$ is stored in the liver, where it remains unchanged or undergoes a limited metabolism because of impaired mixed function oxidase (MFO) activity.[87,88] With protein refeeding, MFO activity increases and leads to an extensive production of other reactive metabolites of AFB$_1$.[88,89] Patients with chronic liver diseases are usually protein – deficient, and thus a similar mechanism may be responsible for the accumulation of unmetabolized AFB$_1$ in the liver.

5. Results of ELISA

In seven liver cancer samples analyzed by using the ELISA procedure, values for a number of AFB$_1$-DNA equivalents ranging from 0.196 to 1.096 ng/AFB$_1$/mg DNA were found. No detectable AFB$_1$ bound to DNA, measured by ELISA as well as assayed by RIA, has been found in one liver sample of cholangiocarcinoma.

Values from the two independent studies (ELISA and RIA) are demonstrated in Table 40 for comparative purposes.

During the last few years, there has been considerable interest in the development of quantitative immunoassay for aflatoxin, and several laboratories have developed antibodies against AFB$_1$ and its metabolites.[90-95] Several works utilizing either mono- or polyclonal antibodies to measure carcinogenic AFB$_1$ adducts in both animals[96,97] and men[98-100] have been

TABLE 40
Level of AFB1 Modification of Human Liver DNA Measured by ELISA

No.	Sex	Age (years)	IH	HBsAg	Alcohol	Ci	Type of tumor	AFB_1 (ng/mg) DNA^a	AFB_1 (ng/g)^b
1	F	5					HC	0.821 ± 0.7580	5.10
2	M	67	+			+	HC	0.406 ± 0.1287	5.42
3	M	69		+	+	+	HC	1.096 ± 0.5275	3.60
4	M	60	+		+	+	HC	0.196 ± 0.771	9.00
5	M	62	+	+		+	HC	0.322 ± 0.1343	3.50
6	M	62		+		+	HC	0.486 ± 0.0665	3.30
7	F	74					ChC	ND	ND
8	M	86			+		HC	0.423 ± 0.0877	3.10

Note: ND = not detectable.
HC = hepatocellular carcinoma.
ChC = cholangiocellular carcinoma.

^{a} Mean of 2 determinations ± ISD.
^{b} Determined by RIA independently.

recently reported. For the potential human liver carcinogen, AFB_1, it has been demonstrated that metabolites M_1, P_1, and AFB_1-guanine[98,99] were found in the urine of people exposed to aflatoxin through dietary contamination, and AFB_1 or its metabolites were detected in the sera of male Japanese subjects.[101]

Some years ago, Booth et al.[102] reported that human liver slices had the ability to activate AFB_1 to its 8,9-oxide, which subsequently bound to DNA. The recent findings of AFB_1-DNA adduct in the liver tissue of liver cancer patients have for the first time demonstrated that a similar process may occur *in vivo* and that AFB_1 binds to human liver DNA of people exposed to aflatoxin environmentally.[74]

It is, however, noteworthy that inhibition of the antibody can, occasionally, take place nonspecifically. At DNA concentrations used in this assay (1 mg/ml), no evidence for such inhibition was found. Moreover, in the liver sample of cholangiocarcinoma, no detectable AFB_1 bound to DNA was found, measured independently either by ELISA or assayed by RIA. This observation argues strongly that antibody inhibition was indeed specific.

Animal studies do indicate that there is an accumulation of persistent AFB_1-DNA adduct, iro-AFB_1-DNA under chronic exposure conditions, providing a rational basis as to why AFB_1-adducts may be detectable in the human liver, maybe even after AFB_1-exposure has ceased.[103]

6. Sources of Aflatoxin

The source of aflatoxin ingested by the patients was not found. In contrast to the developing countries where food available for the population is mostly monotonous, the ascertainment of aflatoxin-contaminated food is easier than in advanced societies with a wide variety of foodstuffs. It also seems unlikely that the population in the Western countries may be exposed to a chronic ingestion of food heavily contaminated by this toxin, as is the case in the tropical areas. However, it cannot be excluded that even low levels of the toxin may be within the carcinogenic range and thus hazardous for man.

The most significant risk of aflatoxin from dietary patterns in the Western countries is most probably presented by food substrates and foodstuffs imported from the tropical areas. A study in Czechoslovakia has shown that 64.9% of 57 strains of *Aspergillus flavus* isolated from imported peanuts, almonds, and crushed coconuts produced AFB_1.[104] However, several surveys of foodstuffs undertaken in Czechoslovakia[105,106] have shown that even several inland cereals and baker's products were contaminated by aflatoxin-producing strain of *A. flavus*. Moreover, aflatoxin-producing strains of *A. flavus* were found in crate boards,[107] with subsequent penetration of the toxin into the food. These findings provide evidence that some strains of *A. flavus* may produce the toxin even under the European climatic conditions.

C. CONCLUSION

Carcinogenesis is a complex process, and identification of any single agent as a primary cause is difficult. To date, the epidemiological evidence in Africa and Asia strongly implicates both aflatoxin and HBV as the major agents in the production of PLC.[71] In the light of the recent study, demonstrating AFB_1 and AFB_1-DNA adduct in the liver tissue of liver cancer patients, it is assumed that besides HBV infection, aflatoxin, though often ignored in the past, must now be considered an important factor in liver carcinogenesis even in the advanced countries of the world.

To clarify this problem concerning the etiological role of HBV infection and aflatoxin in liver carcinogenesis, various efforts have been planned. A vaccination program against HBV in those areas of the world where HBV infection is endemic and the outcome of vaccination may be expected to reduce or eradicate liver cancer.[108,109]

A concurrent program directed to defining better the role of aflatoxin as a human hepatocarcinogen by monitoring human aflatoxin exposure has been proposed.[110] The existing progress in immunological techniques, using specific antibodies against aflatoxin and its metabolic by-products, offers the possibility of determining individual exposure in a large human population.

REFERENCES

1. **Carnaghan, R. B. A.**, Hepatic tumors in ducks fed a low level of toxic groundnut meal, *Nature (London)*, 208, 308, 1965.
2. **Halver, J. E.**, Aflatoxicosis and trout hepatoma, in *Aflatoxin: Scientific Background, Control and Implications*, Goldblatt, L. A., Ed., Academic Press, New York, 1969, 265.
3. **Newberne, P. M. and Wogan, G. N.**, Sequential morphologic changes in aflatoxin B_1 carcinogenesis in the rat, *Cancer Res.*, 28, 770, 1968.
4. **Sato, S., Matsushima, T., Tanaka, N., Sugimura, T., and Takashima, F.**, Hepatic tumors in the guppy *(Lebister reticulatus)* induced by aflatoxin B_1, dimethylnitrosamine, and 2-acetylaminofluorene, *J. Natl. Cancer Inst.*, 50, 765, 1973.
5. **Adamson, R. H., Correa, P., Sieber, S. M., McIntire, K. R., and Dalgard, D. W.**, Carcinogenicity of aflatoxin B_1 in rhesus monkeys: two additional cases of primary liver cancer, *J. Natl. Cancer Inst.*, 57, 67, 1976.
6. **Reddy, J. K., Svoboda, D. J., and Rao, M. S.**, Induction of liver tumors by aflatoxin B_1 in the tree shrews *(Tupaia glis)*, a nonhuman primate, *Cancer Res.*, 36, 151, 1976.
7. **Shank, R. C., Bhamarapravati, N., Gordon, J. E., and Wogan, G. N.**, Dietary aflatoxins and human liver cancer. IV. Incidence of primary liver cancer in two municipal populations of Thailand, *Food Cosmet. Toxicol.*, 10, 171, 1972.
8. **Shank, R. C., Gordon, J. E., Wogan, G. N., and Nondasuta, A.**, Dietary aflatoxins and human liver cancer. II. Aflatoxins in market foods and foodstuffs of Thailand and Hong Kong, *Food Cosmet. Toxicol.*, 10, 61, 1972.
9. **Shank, R. C., Gordon, J. E., Wogan, G. N., Nondasuta, A., and Subhamani, B.**, Dietary aflatoxins and human liver cancer. III. Field survey of rural Thai families for ingested aflatoxins, *Food Cosmet. Toxicol.*, 10, 71, 1972.
10. **Keen, P. and Martin, P.**, The toxicity and fungal infestation of foodstuffs in Swaziland in relation to harvesting and storage, *Trop. Geogr. Med.*, 23, 35, 1971.
11. **Keen, P. and Martin, P.**, Is aflatoxin carcinogenic in man? The evidence in Swaziland, *Trop. Geogr. Med.*, 23, 44, 1971.
12. **Alpert, H. E., Hutt, M. S. R., Wogan, G. N., and Davidson, C. S.**, Association between aflatoxin content of food and hepatoma frequency in Uganda, *Cancer*, 28, 253, 1971.

13. **Peers, F. G. and Linsell, C. A.,** Dietary aflatoxins and liver cancer — a population based study in Kenya, *Br. J. Cancer,* 27, 473, 1973.
14. **Peers, F. G., Gilman, G. A., and Linsell, C. A.,** Dietary aflatoxin and human liver cancer. A study in Swaziland, *Int. J. Cancer,* 17, 167, 1976.
15. **Peers, F. G. and Linsell, C. A.,** Dietary aflatoxins and human primary liver cancer, *Ann. Nutr. Aliment.,* 31, 1005, 1977.
16. **VanRensburg, S. J., Kirsipuu, A., Countinho, L. P., and van der Watt, J. J.,** Circumstances associated with the contamination of food by aflatoxin in a high primary liver cancer area, *S. Afr. Med. J.,* 49, 877, 1975.
17. **LeBreton, E., Frayssinet, C., and Boy J.,** Sur l'apparition d'hepatomes spontanés chez le rat Wistar. Role de las toxine de l'*Aspergillus flavus.* Interet en pathologie humaine et cancerologie experimentale, *C. R. Acad. Sci. Paris,* 225, 784,.1962.
18. **Oettle, A. G.,** The etiology of liver carcinoma in Africa with an outline of the mycotoxin hypothesis, *S. Afr. Med. J.,* 39, 817, 1965.
19. **World Health Organization,** *Environmental Health Criteria 11,* Mycotoxins, WHO, Geneva, 1979, 68.
20. **Masimango, N. T. and Kalengayi, M. M. R.,** Aflatoxins in food and foodstuffs in Zaire, in *Proc. Int. Symp. Mycotoxins,* Cairo, Egypt, September 6-8, 1981, 1983, 431.
21. **Anukarahanonta, T., Chudhabuddhi, C., Temcharoen, P., and Sukroongreung, S.,** Cancer risk from aflatoxins in Thailand, in *Toxigenic Fungi,* Kurata, H. and Ueno, Y., Eds., Elsevier, Amsterdam, 1984, 339.
22. **Carlborg, F. W.,** Cancer, mathematical models and aflatoxin, *Food Cosmet. Toxicol.,* 17, 159, 1979.
23. **Hsieh, D. P. H. and Ruebner, B. H.,** An assessment of cancer risk from aflatoxin B_1 and M_1, in *Toxigenic Fungi,* Kurata, H. and Ueno, Y., Eds., Elsevier, Amsterdam, 1984, 332.
24. **Beasley, R. P.,** Hepatitis B virus as the etiologic agent in hepatocellular carcinoma. Epidemiological considerations, Hepatology, 2, 213, 1982.
25. **Mason, W. S., Halpern, M. S., and Thomas, W.,** Hepatitis B viruses, liver disease and hepatocellular carcinoma, *Cancer Suv.,* 3, 1, 1984.
26. **Deinhardt, F. and Gust, I. D.,** Viral hepatitis, *Bull. W.H.O.,* 60, 661, 1982.
27. **Beasley, R. P., Hwang, L. Y., Lin, C. C., and Chien, C. S.,** Hepatocellular carcinoma and hepatitis B virus, *Lancet,* 2, 1129, 1981.
28. **Tabor, E., Gerety, R. J., Vogel, C. L., Bayley, A. C., Anthony, P. P., Chan, C., and Barker, L. F.,** Hepatitis B virus infection and primary hepatocellular carcinoma, *J. Natl. Cancer Inst.,* 58, 1197, 1977.
29. **Coursaget, P., Maupas, P., Goudeau, A., Chiron, J., Drucker, J., Denis, F., and Diop-Mar, I.,** Primary hepatocellular carcinoma in intertropical Africa: relationship between age and hepatitis B virus etiology, *J. Natl. Cancer Inst.,* 65, 687, 1980.
30. **FuSun Yeh, Chi-Chun Mo, SiLuo, Henderson, B. E., Tong, M. J., and Yu, M. C.,** A serological case-control study of primary hepatocellular carcinoma in Guangxi, China, *Cancer Res.,* 45, 872, 1985.
31. **Lam, K. C., Yu, M. C., Leung, J. W. C., and Henderson, B. E.,** Hepatitis B virus and cigarette smoking: risk factors for hepatocellular carcinoma in Hong Kong, *Cancer Res.,* 42, 5246, 1982.
32. **Chien, M. C., Tong, M. J., Lo, K. J., Lu, J. K., Milich, D. R., Vgas, G. N., and Murphy, B. L.,** Hepatitis B viral markers in patients with primary hepatocellular carcinoma in Taiwan, *J. Natl. Cancer Inst.,* 66, 475, 1981.
33. **Sun, T. and Chu, Y.,** Carcinogenesis and prevention strategy of liver cancer in areas of its prevalence, *J. Cell. Physiol.,* 3 (Suppl.), 39, 1984.
34. **Lin, J. J., Chien Liu, and Svoboda, D. J.,** Long term effects of aflatoxin B_1 and viral hepatitis on marmoset liver. A preliminary report, *Lab. Invest.,* 30, 267, 1974.
35. **Svoboda, D. J., Reddy, J. K., and Chien Liu,** Multinucleate giant cells in livers of marmosets given aflatoxin B_1, *Arch. Pathol.,* 91, 452, 1971.
36. **Wogan, G. N.,** Mycotoxins and other naturally occurring carcinogens, in *Environmental Cancer, Advances in Modern Toxicology,* Kraybill, H. F. and Mehlman, M. A., Eds., Hemisphere Publishing Corporation, Washington, D.C., 1977, 263.
37. **Swenson, D. H., Miller, E. C., and Miller, J. A.,** Aflatoxin B_1-2,3 oxide; evidence for its formation in rat liver *in vivo* and by human microsomes *in vitro, Biochem. Biophys. Res. Commun.,* 60, 1036, 1974.
38. **Lutwick, L.,** Relation between aflatoxin, hepatitis B virus, and hepatocellular carcinoma, *Lancet,* 1, 755, 1979.
39. **Bennett, J. W. and Christensen, S. B.,** New perspectives on aflatoxin biosynthesis, *Adv. Appl. Microbiol.,* 29, 53, 1983.
40. **Enwonwu, C. O.,** The role of dietary aflatoxin in the genesis of hepatocellular cancer in developing countries, *Lancet,* 1, 956, 1984.
41. **Pranco, D., Castaing, D., Bréchot, C., and Morin, J.,** L'aflatoxine B_1 est-elle uncarcinogene hépatique chez l'homme?, *Gastroenterol. Clin. Biol.,* 6, 125, 1982.
42. **Ghadially, F. N.,** Hypertrophy of smooth endoplasmic reticulum in hepatocytes, in *Ultrastructural Pathology of the Cell and Matrix,* Ghadially, F. N., Ed., Butterworths, London, 1982, 368.
43. **Boonchuvit, B. and Hamilton, P. B.,** Interaction of aflatoxin and paratyphoid infections in broiler chickens, *Poult. Sci.,* 54, 1567, 1975.

44. **Hamilton, P. B. and Harris, J. R.,** Interaction of aflatoxicosis with *Candida albicans* infections and other stresses in chickens, *Poult. Sci.,* 50, 906, 1971.
45. **Edds, G. T., Nair, K. P. C., and Simpson, C. F.,** Effect of aflatoxin B_1 on resistance in poultry against cecal coccidiosis and Marek's disease, *Am. J. Vet. Res.,* 34, 819, 1973.
46. **Thaxton, J. P., Tung, H. T., and Hamilton, P. B.,** Immunosuppression in chickens by aflatoxin, *Poult. Sci.,* 53, 721, 1974.
47. **Mohapatra, N. and Roberts, J. F.,** Effects of aflatoxin B_1 on rat peritoneal macrophages and mouse fibroblasts (L-M cells), *Gen. Pharmacol.,* 10, 471, 1979.
48. **Pier, A. C., Fichtner, R. E., and Cysewski, S. J.,** Effects of aflatoxin on the cellular immune system, *Ann. Nutr. Aliment.,* 31, 781, 1977.
49. **Savel, H., Forsyth, B., and Schaeffer, W.,** Effect on aflatoxin B_1 upon phytohemagglutinin-transformed human lymphocytes, *Proc. Soc. Exp. Biol. Med.,* 134, 1112, 1970.
50. **Hahon, N., Booth, J. A., and Stewart, J. D.,** Aflatoxin inhibition of viral interferon induction, *Antimicrob. Agents Chemother.,* 16, 277, 1979.
51. **Newberne, P. M.,** Interaction of nutrients and other factors with mycotoxins, in *FAO/UNEP/USSR, International Training Course,* Moscow, 1985, 33.
52. **Bulatao-Jayme, J., Almero, E. M., Castro, C. A., Jardeleza, T. H., and Salamat, L. A.,** A case control dietary study of primary liver cancer risk from aflatoxin exposure, *Int. J. Epidemiol.,* 11, 112, 1982.
53. **Newberne, P. M.,** Interaction of nutrients and other factors with mycotoxins, in *FAO/UNEP/USSR, International Training Course,* Moscow, 1985, 50.
54. **Neurath, G. B. and Schreiber, O.,** Investigations on amines in the human environment, in *IARC, N-Nitroso Compounds in the Environment,* Bogowski, E. A., Ed., IARC Scientific Publications, Lyon, 1975, 211.
55. **Angsubhakorn, S., Bhamarapravati, N., Romruen, K., and Sahaphong, S.,** Enhancing effects of dimethylnitroamine on aflatoxin B_1 hepatocarcinogenesis in rats, *Int. J. Cancer,* 28, 621, 1981.
56. **Higginson, J.,** The geographical pathology of primary liver cancer, *Cancer Res.,* 23, 1624, 1963.
57. **Higginson, J. and Svoboda, D. J.,** Primary carcinomas of the liver as a pathologist's problem, in *Pathology Annual,* Sheldon, C. S., Ed., Butterworths, London, 1970, 61.
58. **Manderson, W. G., Patrick, R. S., and Peters, E. E.,** Incidence of primary carcinoma of the liver in the west of Scotland between 1949 — 1965, *Gut,* 9, 480, 1968.
59. **MacSween, R. N. M.,** A clinicopathological review of 100 cases of primary malignant tumours of the liver, *J. Clin. Pathol.,* 27, 669, 1974.
60. **Glenert, J.,** Primary carcinoma of the liver: a post mortem study of 104 cases, *Acta Pathol. Microbiol. Scand.,* 53, 50, 1961.
61. **Ohlsson, E. G. H. and Norden, J. G.,** Primary carcinoma of the liver. A study of 121 cases, *Acta Pathol. Microbiol. Scand.,* 64, 430, 1965.
62. **Elkingotn, S. G., McBrien, D. J., and Spencer, H.,** Hepatoma in cirrhosis, *Br. Med. J.,* 2, 1501, 1963.
63. **Patton, R. B. and Horn, R. C., Jr.,** Primary liver carcinoma: autopsy study of 60 cases, *Cancer,* 17, 757, 1964.
64. **Burnett, R. A., Patrick, R. S., Spilg, W. G. S., Buchanan, W. M., and MacSween, R. N. M.,** Hepatocellular carcinoma and hepatic cirrhosis in the west of Scotland: a 25-year necropsy review, *J. Clin. Pathol.,* 31, 108, 1978.
65. **Gall, E. A.,** Primary and metastatic carcinoma of the liver. Relationship to hepatic cirrhosis, *Arch. Pathol.,* 70, 226, 1960.
66. **Lee, F. I.,** Cirrhosis and hepatoma in alcoholics, *Gut,* 7, 77, 1966.
67. **MacSween, R. N. M. and Scott, A. R.,** Hepatic cirrhosis: a clinicopathological review of 520 cases, *J. Clin. Pathol.,* 26, 936, 1973.
68. **Trichopoulos, D., Tabor, E., and Gerety, R. J.,** Hepatitis B and primary hepatocellular carcinoma in a European population, *Lancet,* 2, 1217, 1978.
69. **Bassendine, M. F., Chadwick, R. G., Lyssiotis, T., Thomas, H. C., Sherlock, S., and Cohen, B. J.,** Primary liver cell cancer in Britain a viral oetiology?, *Br. Med. J.,* 1, 166, 1979.
70. **Yarrish, R. L., Werner, B. G., and Blumberg, B. S.,** Association of hepatitis B virus infection with hepatocellular carcinoma in American patients, *Int. J. Cancer,* 26, 711, 1980.
71. **Lutwick, L.,** Relation between aflatoxin, hepatitis B virus, and hepatocellular carcinoma, *Lancet,* 1, 755, 1979.
72. **Shikata, T., Uzawa, T., Yoshimura, N., Akutsaka, T., and Yamazaki, S.,** Staining methods of Australia antigen in paraffin sections, *Jpn. J. Exp. Med.,* 44, 25, 1974.
73. **Tanaka, K., Mori, W., and Suwa, K.,** Victoria blue — nuclear fast red stain for HBs antigen detection in paraffin section, *Acta Pathol. Jpn.,* 31, 93, 1981.
74. **Tursi, F., Dvořáčková, I., and Garner, R. C.,** Human and rat exposure to aflatoxin B_1 at the macromolecular level measured by immunoassay, *Carcinogenesis,* in print.
75. **Stanton, C. A., Chow, F. L., Phillips, B. H., Grover, P. L., Garner, R. C., and Martin, C. N.,** Evidence for N-(deoxyguanosin-8yl)-1-aminopyrene as a major DNA adduct in female rats treated with 1-nitopyrene, *Carcinogenesis,* 6, 535, 1985.

76. **Martin, C. N., Garner, R. C., Garber, J. V., Whittle, H. C., Sizaret, R., and Montesano, R.,** An enzyme-linked immunosorbent procedure for assaying aflatoxin B_1, in *Monitoring Human Exposure to Carcinogenic and Mutagenic Agents,* Berlin, A., Draper, M., Hemminki, K., and Vaino, H., Eds., IARC Scientific Publications, Lyon, 1984, 313.
77. **Deinhardt, P. and Gust, I. D.,** Viral hepatitis, *Bull. W.H.O.,* 60, 661, 1982.
78. **Tabor, E., Gerety, R. J., Vogel, C. L., Bayley, A. C., Anthony, P. P., Chan C., and Barker, L. F.,** Hepatitis B virus infection and primary hepatocellular carcinoma, *J. Natl. Cancer Inst.,* 58, 1197, 1977.
79. **Woodfield, D. G., Endo, Y., and Matsuhashi, T.,** Primary liver cancer, afla l fetoprotein and hepatitis B antigen in Papua, New Guinea, *Aust. N.Z. J. Med.,* 4, 3, 1974.
80. **Sung, J. L.,** Hepatitis and hepatoma in Taiwan, *Jpn. J. Gastroentrol.,* 70, 977, 1973.
81. **Fu-Sun Yeh, Chi-Chun Mo, Si Luo, Henderson, B. E., Tong, M. J., and Mimi C. Yu,** A serological case-control study of primary hepatocellular carcinoma in Guangxi, China, *Cancer Res.,* 45, 872, 1985.
82. **Burdett, W. J.,** Neoplasms of the liver, in *Diseases of the Liver,* Schiff, L., Ed., Lippincott, Philadelphia, 1975, 1051.
83. **Wray, B. B. and Hayes, A. W.,** Aflatoxin B_1 in the serum of a patient with primary hepatic carcinoma, *Environ. Res.,* 22, 400, 1980.
84. **Phillips, D. L., Yourtee, D. M., and Searles, S.,** Presence of aflatoxin B_1 in human liver in the United States, *Toxicol. Appl. Pharmacol.,* 36, 403, 1976.
85. **Jukes, T. H.,** Corn and peanuts, *Nature (London),* 271, 499, 1978.
86. **Dalezios, J. and Wogan, G. N.,** Metabolism of aflatoxin B_1 in rhesus monkey, *Cancer Res.,* 32, 2297, 1972.
87. **Lamplugh, S. M. and Hendrickse, R. G.,** Aflatoxin and kwashiorkor, *Afr. Health,* 5, 20, 1983.
88. **Enwonwu, C. O.,** Kwashiorkor: further thoughts on the aflatoxin theory, *Afr. Health,* 6, 18, 1984.
89. **Enwonwu, C. O.,** The role of dietary aflatoxin in the genesis of hepatocellular cancer in developing countries, *Lancet,* 1, 956, 1984.
90. **Chu, F. S.,** Immunoassay for analysis of mycotoxins, *J. Food Prot.,* 47, 562, 1984.
91. **Langone, J. J. and van Vunakis, H.,** Aflatoxin B_1: specific antibodies and their use in radioimmunoassay, *J. Natl. Cancer Inst.,* 56, 591, 1976.
92. **Lawellin, D. W., Grant, D. W., and Yoyce, B. K.,** Enzyme-linked immunosorbent analysis of aflatoxin B_1, *Appl. Environ. Microbiol.,*34, 94, 1977.
93. **Pestka, J. J., Gaur, P. K., and Ch, F. S.,** Quantitation of aflatoxin B_1 and aflatoxin B_1 antibody by an enzyme-linked immunosorbent microassay, *Appl. Environ. Microbiol.,* 40, 1027, 1980.
94. **Gaur, P. K., Lau, P., Pestka, J. J., and Chu, F. S.,** Production and characterization of aflatoxin B_2a antiserum, *Appl. Environ. Microbiol.,* 41, 478, 1981.
95. **Harder, W. O. and Chu, F. S.,** Production and characterization of antibody against aflatoxin M_1, *Experientia (Basel),* 35, 1104, 1979.
96. **Skipper, P. L., Hutchins, D. H., Turesky, R. J., Sabbioni, G., and Tannenbaum, S. R.,** Carcinogen binding to serum albumin, *Proc. Am. Assoc. Cancer Res.,* 26, 90, 1985.
97. **Wild, C. P., Garner, R. C., Montesano, R., and Tursi, F.,** Aflatoxin B_1 binding to plasma albumin and liver DNA upon chronic administration to rats, *Carcinogenesis,* 7, 853, 1986.
98. **Groopman, J. D., Donahue, P. R., Ihu, J., Chen, J., and Wogan, G. N.,** Aflatoxin metabolism in humans: detection of metabolites and nucleic acid adducts in urine by affinity chromatography, *Proc. Natl. Acad. Sci. U.S.A.,* 82, 6492, 1985.
99. **Autrup, H., Bradley, K. A., Shamsuddin, A. K. M., Wakhisi, J., and Wasunna, A.,** Detection of putative adduct with fluorescence characteristics identical to 2,3-dihydro-2-(7'-guanyl)-3 hydroxyaflatoxin B_1 in human urine collected in Murang'a district Kenya, *Carcinogenesis,* 4, 1193, 1983.
100. **Sun, T., Wu, Y., and Wu, S.,** Monoclonal antibody against aflatoxin B_1 and its potential applications, *Chin. J. Oncol.,* 5, 401, 1983.
101. **Tsuboi, S., Nakagawa, T., Tomita, M., Sea, T., Ono, H., Kawamura, K., and Iwamura, N.,** Detection of aflatoxin B_1 in serum samples of male Japanese subjects by radioimmunoassay and high performance liquid chromatography, *Cancer Res.,* 44, 1231, 1984.
102. **Booth, S. C., Bosenberg, H., Garner, R. C., Hertzog, P. J., and Norpoth, K.,** The activation of aflatoxin B_1 in liver slices and in bacterial mutagenicity assays using livers from different species including man, *Carcinogenesis,* 2, 1063, 1981.
103. **Croy, R. G. and Wogan, G. N.,** Temporal patterns of covalent DNA adducts in rat liver after single and multiple doses of aflatoxin B_1, *Cancer Res.,* 41, 197, 1981.
104. **Jesenská, Z., Poláková, O., Polster, M., Kosková, L., and Vaszilková, A.,** Potential producers of aflatoxin in working environment, *Zentralbl. Bakteriol. Mikrobiol. Hyg.,* 172, 382, 1981.
105. **Polster, M. and Tichá, J.,** Toxinogenita plísní isolovanych z cerealií a výrobků z nich, *Acta Hyg. Epidemiol. Microbiol., IHE, Czechoslovakia,* 12, 27, 1978.
106. **Matyášová, J. and Turek, B.,** Výskyt Aspergillus flavus v potravinách našeho trhu, *Acta Hyg. Epidemiol. Microbiol. I.H.E. Czech.,* 12, 36, 1978.

107. **Weberová, M. and Polster, M.,** Výskyt aflatoxinu cv celulosových obalech pro potraviny napadených *Aspergillus flavus, Acta Hyg. Epidemiol. Microbiol. I.H.E. Czech.,* 12, 31, 1978.
108. **Maupas, P., Goudeau, A., Coursaget, P., Drucker, J., and Bagros, P.,** Immunization against hepatitis B in man, *Lancet,* 1, 1367, 1976.
109. **Maupas, P., Goudeau, A., Coursaget, P., Drucker, J., and Bagros, P.,** Hepatitis B vaccine: efficacy in high-risk settings, a two-year study, *Intervirology,* 10, 196, 1978.
110. **Garner, C. R., Rydler, R., and Montesano, R.,** Monitoring of aflatoxins in human body fluids and application to field studies, *Cancer Res.,* 45, 922, 1985.

Chapter 5

POTENTIAL PROFESSIONAL RISK DUE TO AFLATOXIN EXPOSURE VIA THE RESPIRATORY ROUTE

I. AFLATOXIN IN RESPIRABLE CORN DUST PARTICLES

Aflatoxin-contaminated food was considered the main health risk of human exposure to aflatoxin, and a great number of studies have been directed to this problem. Little attention, however, has been paid to the possible health risks associated with the inhalation of aflatoxin contaminated dust, particularly in men professionally handling contaminated grain, seed, and various food and feed substrates.

Until 1980, only three available studies dealing with aflatoxin exposure via the respiratory route were reported. The first report[1] concerned a group of 60 to 70 Dutch employees, working between 1961 and 1969 in a plant for extracting oil from linseeds and peanuts. During all phases of the process great quantities of dust were created in the work areas. In 1968, cases of cancer among this group of workers were noted. In the same year, the presence of aflatoxin in the residue cakes from peanuts was detected, the toxin containing levels of AFB_1 ranging from 75 to 1320 µg/kg. Aflatoxins were then also identified in settled and aerosol dusts at various work places at a concentrations of ca. 250 µg/kg dust. It was estimated that respiratory exposure for the workers ranges from a minimum of 0.04 µg/kg/week to a maximum of 2.5 µg/kg/week.

An epidemiological study was then carried out which covered a total of 12.5 years with a cohort of 55 workers. Eleven cancers of various form occurred in this group of workers, the majority of which affected the respiratory tract. Cancers consisted of bronchial carcinomas (four cases), and carcinomas of the gall bladder, gastrointestinal tract, maxillary sinus, and prostate. Reticulum cell or anaplastic carcinoma involving the cervical lymph nodes and pleural mesothelioma were also observed. It has been concluded that there were strong indications of the influence of an occupational carcinogen, based in part on the observation of cancer incidence in the exposed group of workers compared to the matched control group and the unusual variety of malignant tumors.

An extensive epidemiological study was then carried out up to 1980 which revealed further cancer cases in this exposed group.[2] In the entire study period between 1963 to 1980, there were 16 deaths due to cancer in the aflatoxin-exposed group and seven in the comparison group. Of particular interest was the occurrence of seven deaths due to tumors of the respiratory system, while no primary liver cancer was found among the aflatoxin-exposed workers. Two individuals died as a result of a nonspecific liver disease.

In 1976, colon carcinoma in two research biochemists who had worked with purified aflatoxin for research purposes was reported.[3] One worker had purified aflatoxin from 1962 to 1964 by scraping it from chromatographic plates and developed symptoms in 1971, at 42 years of age. The second had done this work for 12 months between 1969 and 1970 and developed symptoms in 1972, at 28 years of age.

In the same year, two chemical engineers, who had worked on a method for sterilizing Brazilian peanut meal contaminated with *A. flavus,* were reported to have died from alveolar cell carcinoma.[4] One worker, who died at 68 years of age, had done this work for 3 months and developed the first symptoms of the lung disease 3 months later. The second worker died 3 years ago, and no clinical details were available for him.

The significance of airborne dust in the workplace has become increasingly apparent, and several epidemiological studies relating to occupational exposure to aflatoxin have been reported since 1980.

In the U.S., aflatoxin contamination of certain agricultural crops, such as peanuts, corn, and cottonseed, presents a problem particularly in the southeastern states where climatic conditions are conducive to fungal growth. Corn dust collected in the field during the corn harvest in Georgia, was observed to be contaminated with average amounts of AFB_1 of 138 ppb[5] and 12.7 to 142.8 ppb.[6] Airborne dust collected at a grain elevator in Georgia during a routine handling averaged 191.4 ppb AFB_1.[5]

Sorenson et al.[7] examined bulk grain dust from grain terminals in the Superior-Duluth, Minnesota. Although aflatoxin could not be detected in most of the samples, one corn sample contained 130 ppb AFB_1. When the components of this sample were separated according to their aerodynamic diameter and analyzed for aflatoxin, it was found that particles of diameters 7 to 11 μm and <7 μm had much higher levels (695 to 1814 ppb) of aflatoxin than the sample as a whole.

Although the attempts to find aflatoxin-contaminated dust in the northern states of the U.S. were mostly negative,[6,8] AFB_1 has recently been isolated in dust collected from grain elevator in central Illinois at concentrations ranging from nondetectable up to 3.5 ppb.[9] Dust generated during the corn harvest in England was found to be composed mostly of fungus spores,[10] and Wicklow and Shotwell[11] found AFB_1 levels as high as 84,000 ppb for conidia in toxigenic strains of *A. flavus*. Since the average diameter of *A. flavus* conidia is 3.5 μm, the chance of deposition within the lungs and subsequent exposure to the toxin is significantly high.

II. AFLATOXIN IN LUNG TISSUE OF EXPOSED WORKERS

The first work relating to the evidence of aflatoxin in the lung tissue of a patient exposed to aflatoxin inhalation during his work and who died of lung carcinoma was reported in 1976.[4] Subsequently, other cases of lung diseases associated with the detection of aflatoxin in lung tissue have been described.[12,13] With regard to the rarity of such cases in literature, the cases are presented in detail.

A. CASE REPORTS
Patient 1 — A 68-year-old chemical engineer worked for 3 months on various methods for the sterilization of Brazilian peanut meal which was heavily contaminated by *A. flavus*. Three months after finishing this work, he became ill with a high fever and began to expectorate white sputum. An X-ray examination showed cavities in the left lower lobe of the lung. At first the process was considered to be a case of tuberculosis and later a mycotic disease. Two months later, further lesions developed in both his lungs. He died 11 months after the onset of the illness. The autopsy revealed heavy lungs diffusely infiltrated with firm yellow-white or reddish lesions (Figure 74). A histological investigation revealed bands of fibrous tissue and scars scattered throughout the lung parenchyma. The interalveolar septae were well preserved, and the alveoli were lined with high cylindrical or cuboidal epithelial cells. Mitotic figures were rare (Figures 75, 76). A bacteriological investigation of the lungs was negative. A sample of the lung was taken for a chemical investigation. Thin layer chromatography (TLC) of the lung sample showed a blue fluorescence spot in 365 nm UV light similar to that of the commercial sample of AFB_1 (Calbiochem, California), the same color change as the standard AFB_1 when treated with 50% H_2SO_4 and R_F value identical to that of the commercial AFB_1.

Patient 2 — A 59-year-old cattle breeder, a smoker of 20 cigarettes per day, fell ill with pneumonia in 1972. A bacteriological examination of his sputum revealed *Pneumococci* and *Staphylococcus aureus,* but already at this time, precipitins to *Aspergillus* antigens in the patient's serum were positive. His condition improved after an antibiotic therapy, but an inflammatory infiltration in the upper right lobe remained. In 1975 a relapse of pneumonia developed, and an X-ray examination showed a cavity in the right apex. Cultures tested for *Aspergillus* were negative. In 1977, the patient was admitted to a hospital with dyspnea and fever. The X-ray investigation revealed a cavity filled with a fungus ball in the right apex (Figure

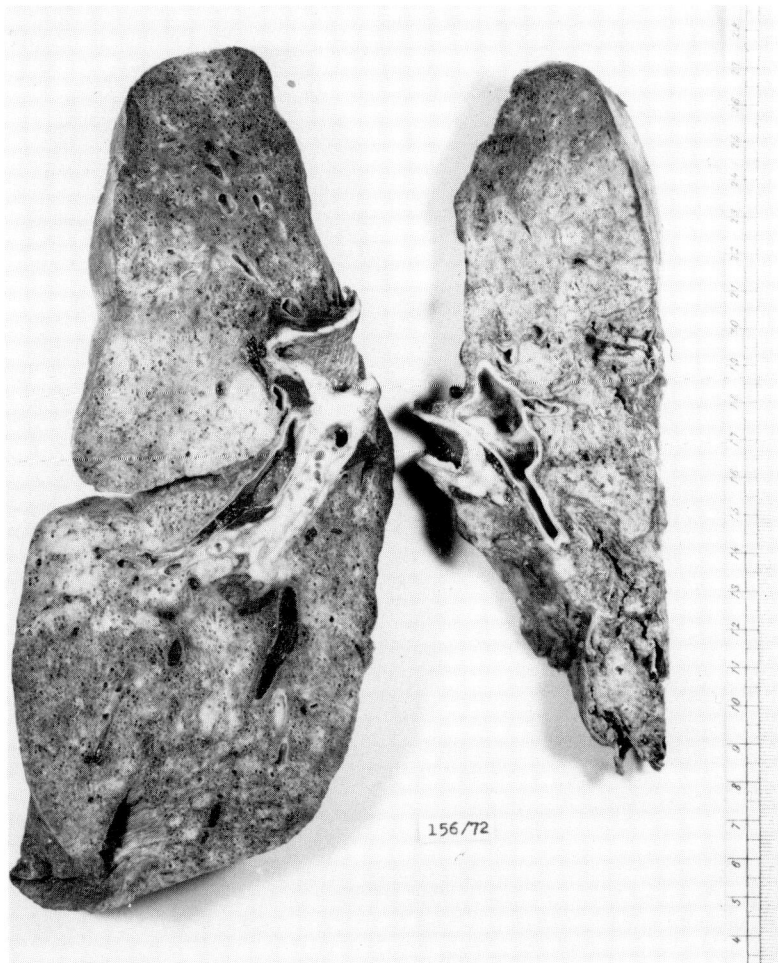

FIGURE 74. Diffuse pulmonary alveolar cell carcinoma replacing most of the lobes of both lungs.

77). The patient was treated with *Fungicidin*, but an allergic reaction appeared and the antifungal therapy had to be stopped. The patient died of cardiorespiratory insufficiency in January 1978. At the autopsy a cavity was found in the apex of the right lung filled with an aspergilloma 3 cm in diameter and yellow-green in color (Figure 78). The wall of the cavity was thick, infiltrated by whitish granular tissue. The remaining lung parenchyma was of stiff consistency; the bronchi contained purulent material. Histologically the wall of the cavity was composed of fibrous collagenous tissue in which rests of bronchial cartilages and plugs of squamous cell carcinoma were found (Figure 79). In the surrounding lung parenchyma, chronic pneumonia was present. No metastases of the tumor were found. The mycetoma consisted of a conglomerate of fungal elements, arranged in concentric layers, resembling the zonal growth observed in cultures of *Aspergillus* on solid media (Figure 80). The hyphae were surrounded by amorphous eosinophilic material very likely resembling the Hoeppli-Splendore phenomenon (Figure 81). On the border, many swollen degenerated and macerated hyphae were present (Figure 82). A part of mycetoma was inoculated on the Czapek-Dox's medium and cultivated at 25°C. After 4 d, dark brown-green and light yellow-green colonies were observed. An attempt at isolating the different colonies was unsuccessful, and the monosporic isolation could not be performed for technical reasons. With regard to a suspicion of the presence of *A. flavus,* a chemical analysis of the culture

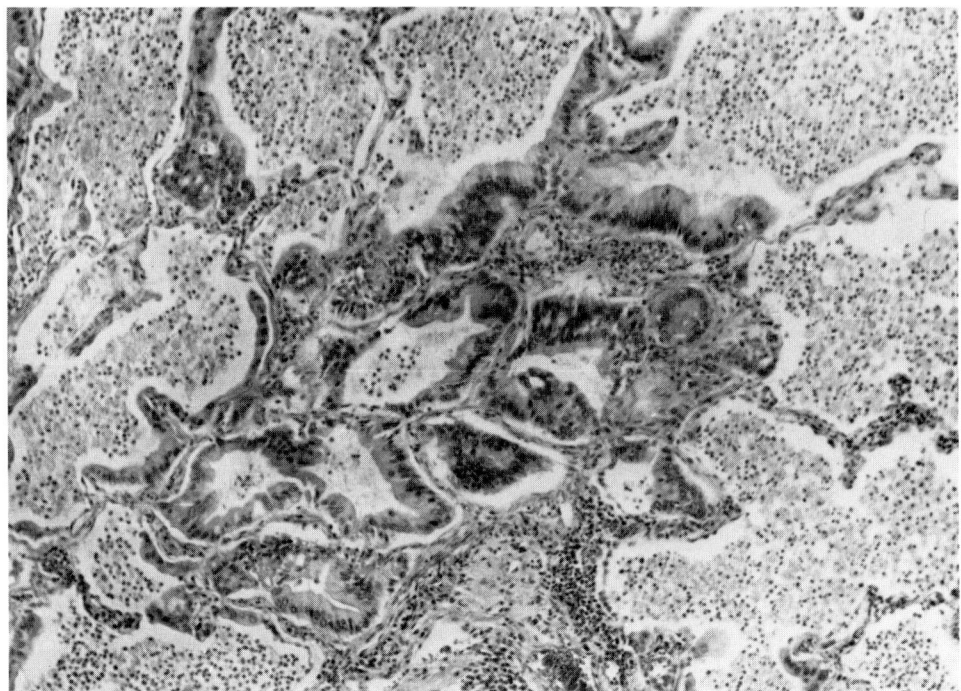

FIGURE 75. Alveolar cell lung cancer showing clumps of tumor cells attached to the walls of alveoli (hematoxylin and eosin, ×150).

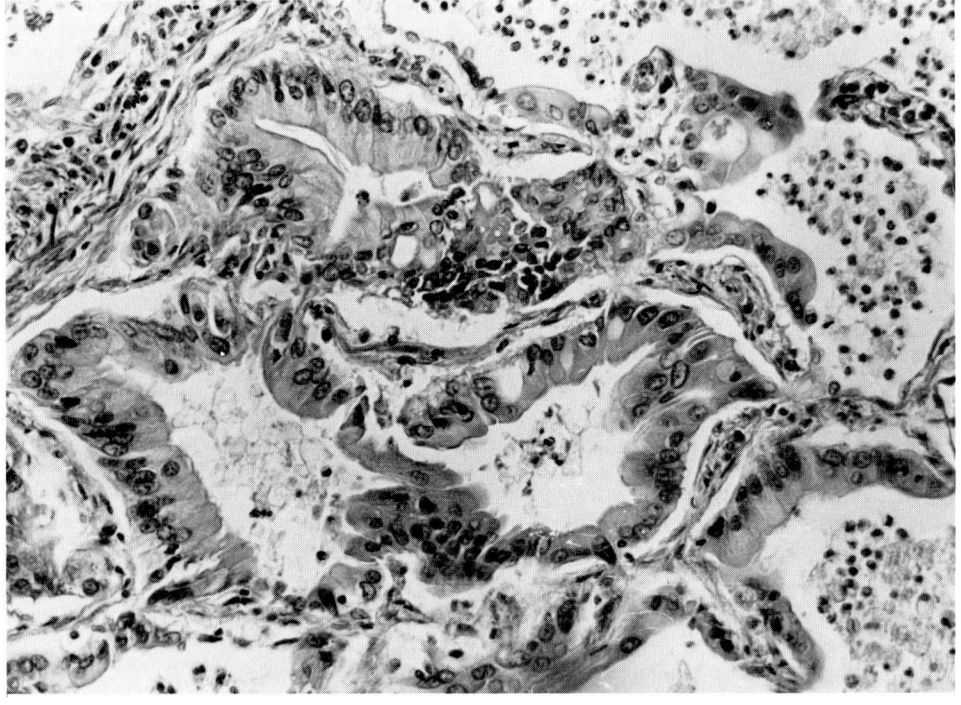

FIGURE 76. Alveolar cell lung carcinoma showing high cylindrical cells lining the walls of the alveoli. Some of the tumor cells contain mucous vacuoles (hematoxylin and eosin, ×230).

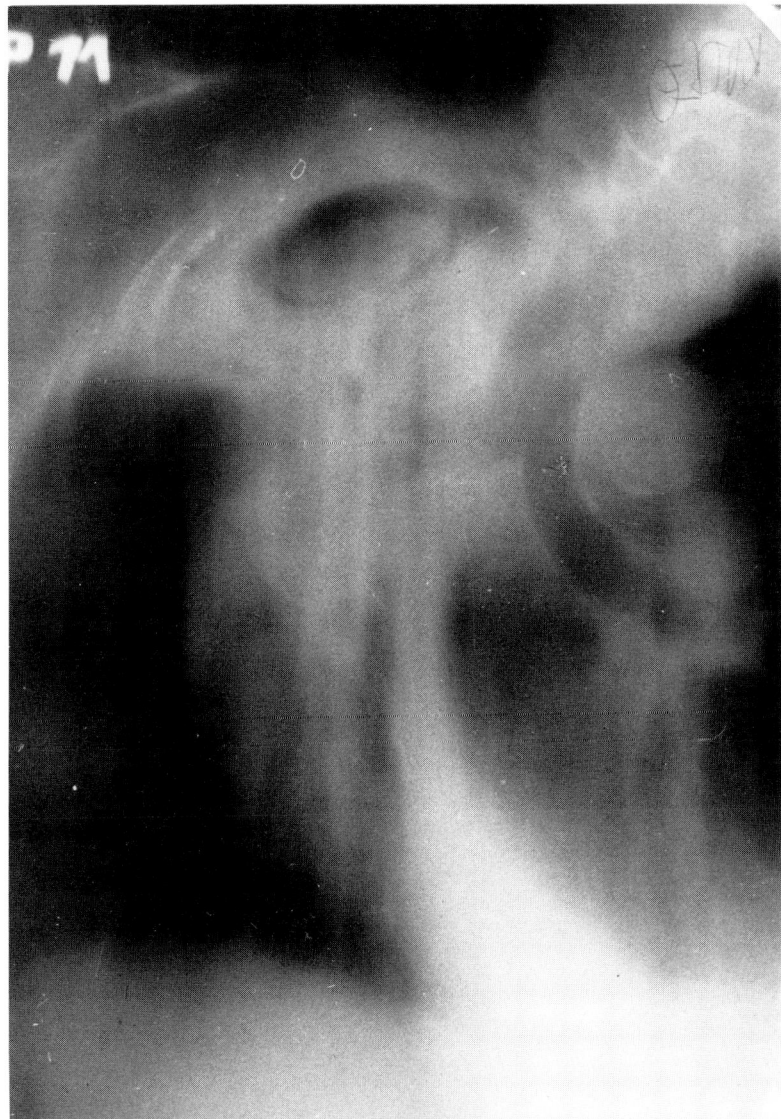

FIGURE 77. Tomogram: a fungus ball in the cavity of the right pulmonary apex.

medium was carried out. The spores were cultivated on crushed crackers over 14 d at 25°C. The culture was then sterilized and dried at 60°C for 1 hour and three times extracted by 30 ml of acetone. The filtered extract was evaporated, and the residue was dissolved in 3 ml of chloroform and tested on Silufol with the standard AFB_1 (USOL, Prague) in eluting system chloroform-acetone (9:1, v/v) in the first and 90% ethanol in the second direction. In both cases, spots with blue fluorescence in 365 nm UV light were found, and the same color change observed when treated with 20% H_2SO_4 like that of the standard AFB_1. The R_F value was identical to that of the commercial sample of AFB_1. Samples of lung and liver tissue were taken at the autopsy and sent to the Mycotoxin Research Laboratory in Bilthoven, Netherlands, for a chemical analysis which was carried out by the two-dimensional chromatographic method.[14] The results of the analysis proved the presence of AFB_1 in the amount of 2,640 μg/kg in the lung and 550 μg/kg in the liver sample.

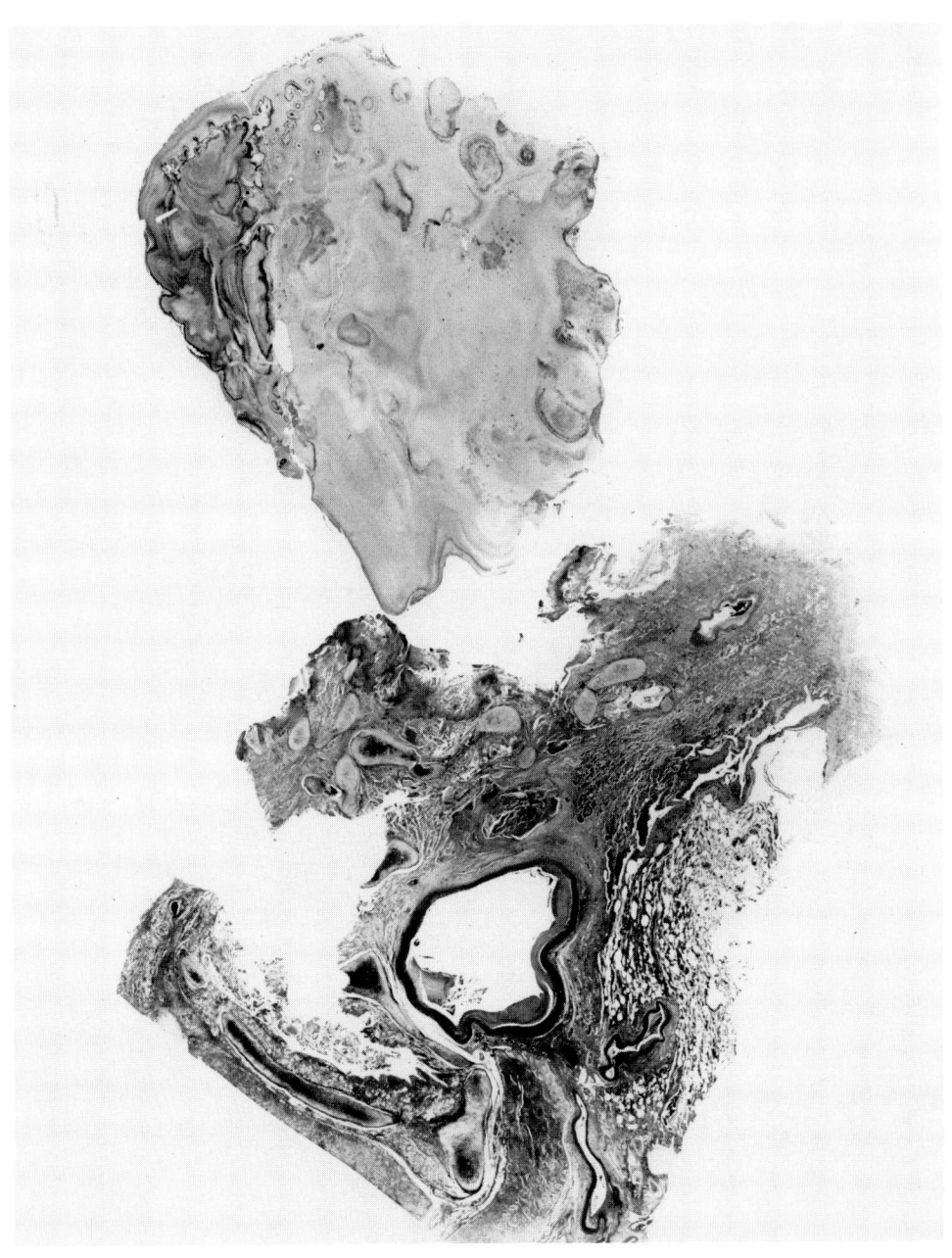

FIGURE 78. A cavity with an aspergilloma (histotopogram, ×5).

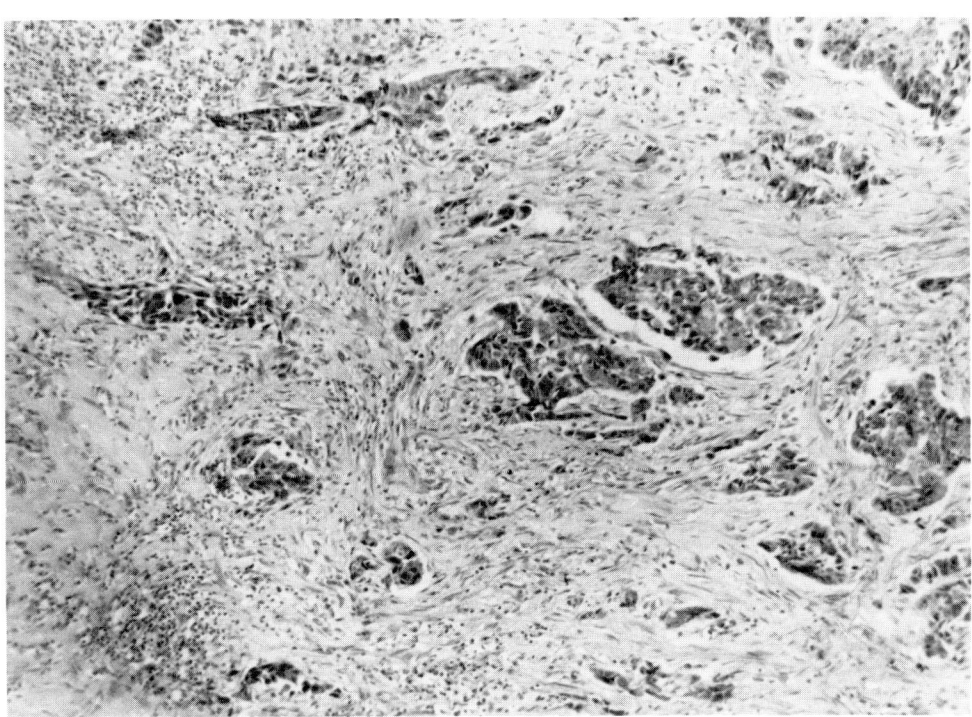

FIGURE 79. Plugs of squamous cell cancer infiltrating the wall of the mycetoma cavity (hematoxylin and eosin, ×130).

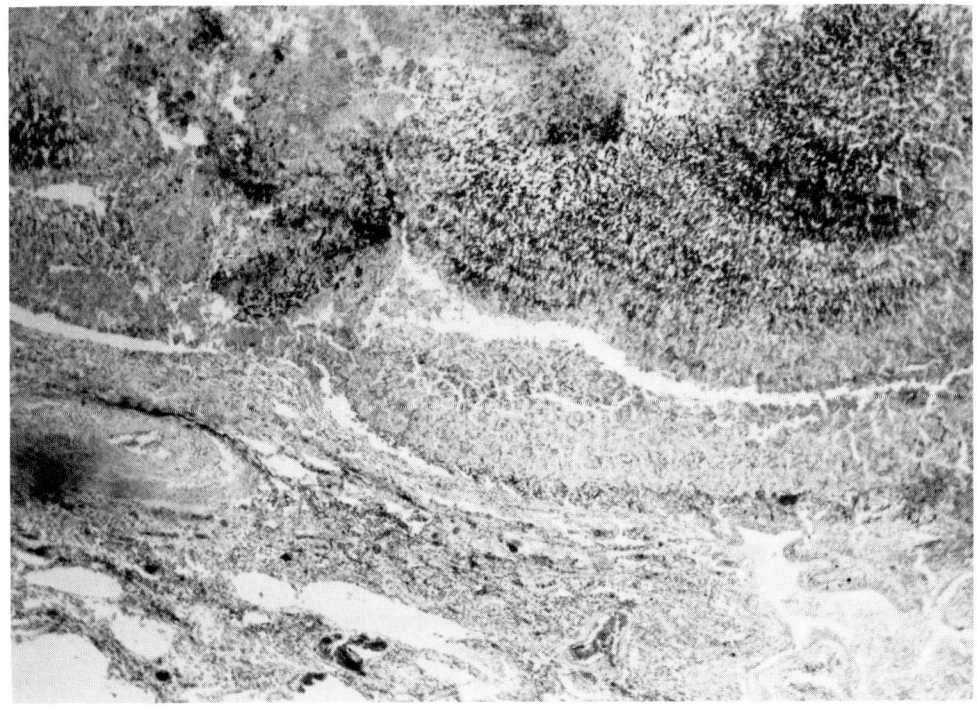

FIGURE 80. The layering within the aspergilloma resembling the zonal growth (Grocott, ×80).

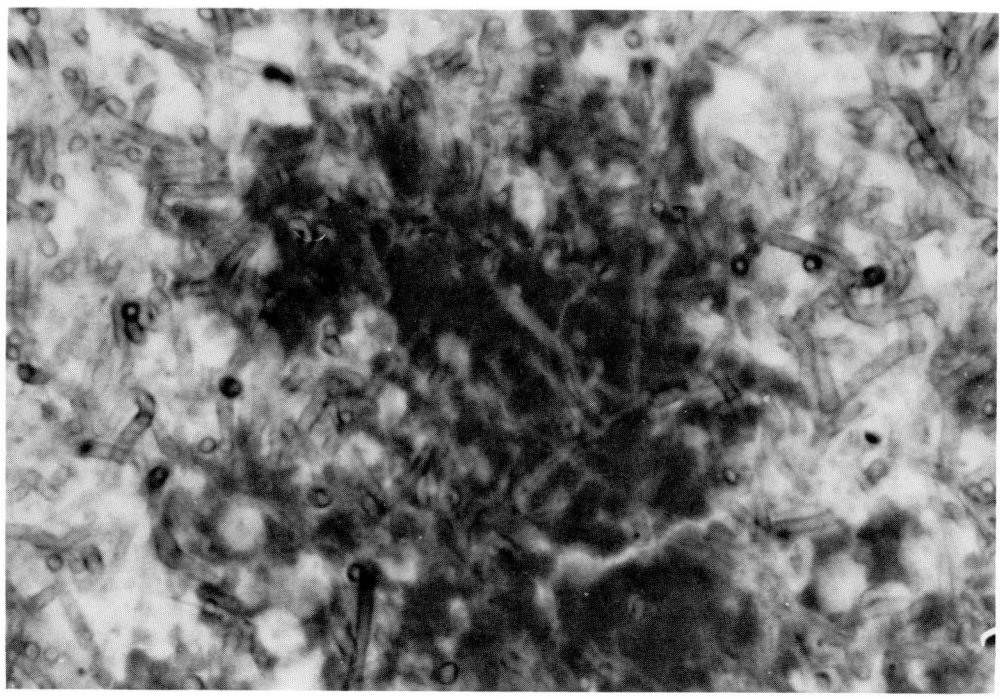

FIGURE 81. Homogenous eosinophilic material surrounding hyphae of *Aspergillus* representing very likely Hoeppli-Splendore phenomenon (hematoxylin and eosin, ×400).

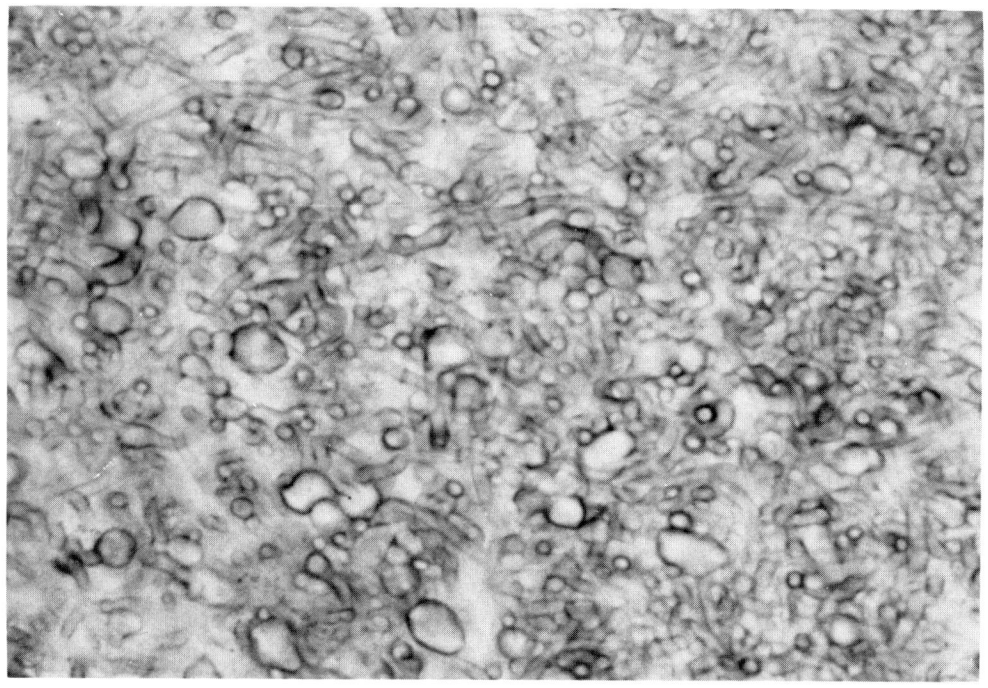

FIGURE 82. Swollen degenerated and macerated hyphae of *Aspergillus* (hematoxylin and eosin, ×400).

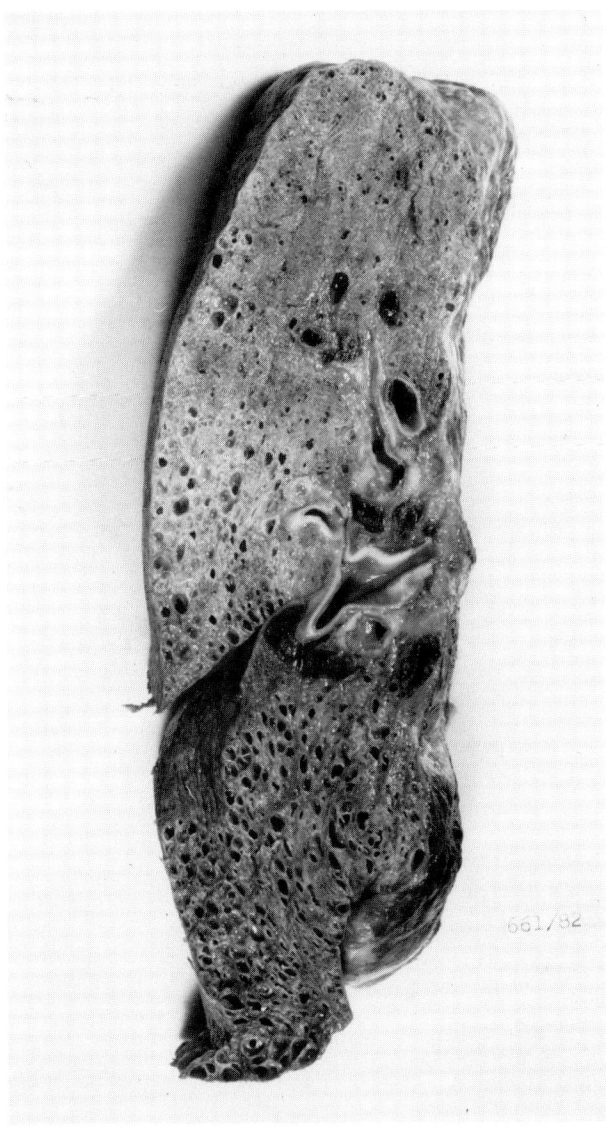

FIGURE 83. Lung cut surface showing large air spaces giving rise to a spongy pattern.

Patient 3 — A 60-year-old cattle breeder, nonsmoker, developed "bronchopneumonia" in 1981. His condition improved after an antibiotic therapy, but dyspnea persisted. One year later, his chest roentgenogram revealed diffuse interstitial process in both lungs. The patient was treated with corticosteroids and died of respiratory insufficiency 2 years after the first symptoms had appeared. At the autopsy, firm lungs with a granular surface were found. The parenchyma was gray in color with numerous small cavities giving rise to a spongy pattern (Figure 83). In the apex of the right lung, a cavity of 13 mm in diameter filled with yellow granular material was present. A histological examination of the lungs revealed diffuse interstitial fibrosis. The interalveolar septa were thickened, formed of collagenous tissue and infiltrated with mononuclear cells. The alveoli were irregularly deformed, some of them were lined with high cuboidal

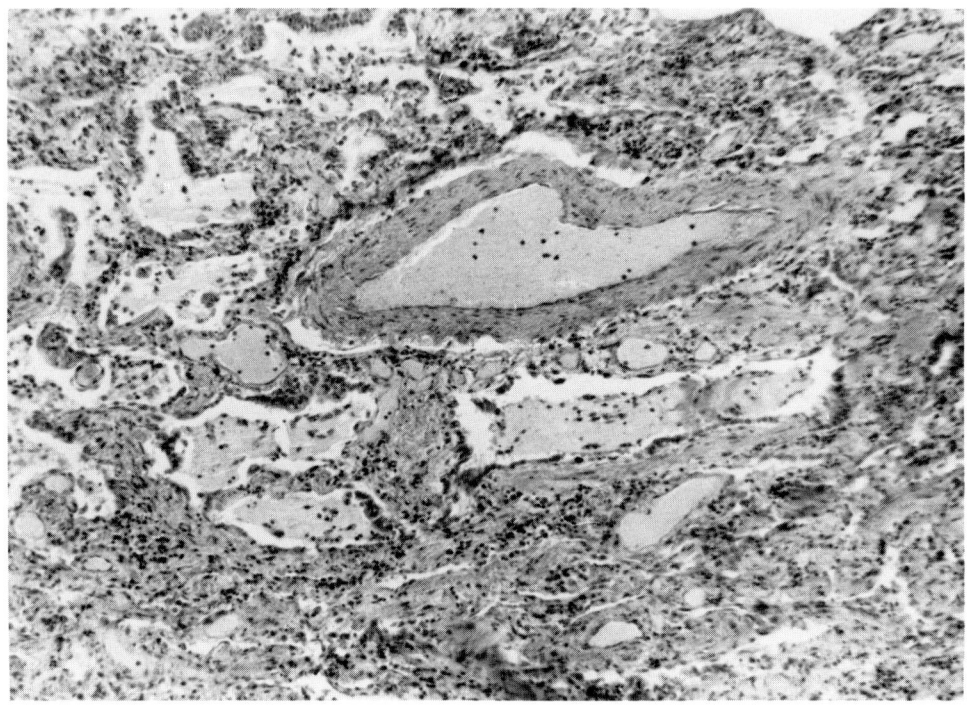

FIGURE 84. Advanced interstitial fibrosis with irregular shape and size of the alveoli. There is a moderate interstitial chronic inflammatory cell infiltration (hematoxylin and eosin, ×90).

cells (Figure 84). Areas of squamous cell metaplasia of the bronchiolar epithelium were found. The mycetoma found in the right apex of the lung was composed of a conglomerate of fungal elements with septate branching hyphae morphologically resembling those of the *Aspergillus* (Figure 85). A cultivation for fungi failed. A serological investigation of the patient's serum for precipitins to fungal antigens gave negative results. In a lung sample taken at the autopsy and assayed by radioimmunoassay (RIA), AFB_1 at the level of 19.9 ng/g was found.

Patient 4 — A 63-year-old woman who worked over 35 years as a weaver in a textile factory developed an irritating cough and dyspnea 7 years before her death. She had numerous relapses of the respiratory illness accompanied with dyspnea. An X-ray examination 3 years before her death revealed irregular interstitial patchy shadowing over both her lung fields. The dyspnea became worse, often accompanied with fits of unconsciousness. The patient died with the following clinical diagnosis: diffuse lung fibrosis and occupational lung damage. The autopsy revealed firm lungs with numerous air spaces separated by areas of fibrosis. A histological investigation showed an irregular shape and size of the alveoli; some of them were lined with cuboidal cells and filled with proteinaceuos exudate and histiocytes (Figure 86). The interalveolar septa were thicker, formed of fibrous tissue infiltrated with varying number of lymphocytes and plasma cells. The bronchioles were dilated and often lined with squamous metaplastic epithelium. Bacteriological and mycological examinations of the lung were negative. A fresh lung sample of 100 g was taken at the autopsy and assayed for aflatoxin by RIA. In the lung sample, AFB_1 at the level of 54 ng/g was detected.

Patient 5 — A 59-year-old stock keeper was taken ill with "influenza" pneumonitis followed by dyspnea and malaise. His condition improved after 2 months, and he continued his work. One and a half years later he complained of dyspnea. An X-ray examination of the lungs revealed irregular interstitial shadowing over both his lungs and enlargement of the hilar lymph nodes. Clinically, a lung cancer was suspected, and the patient was admitted to a radiologic clinic. He

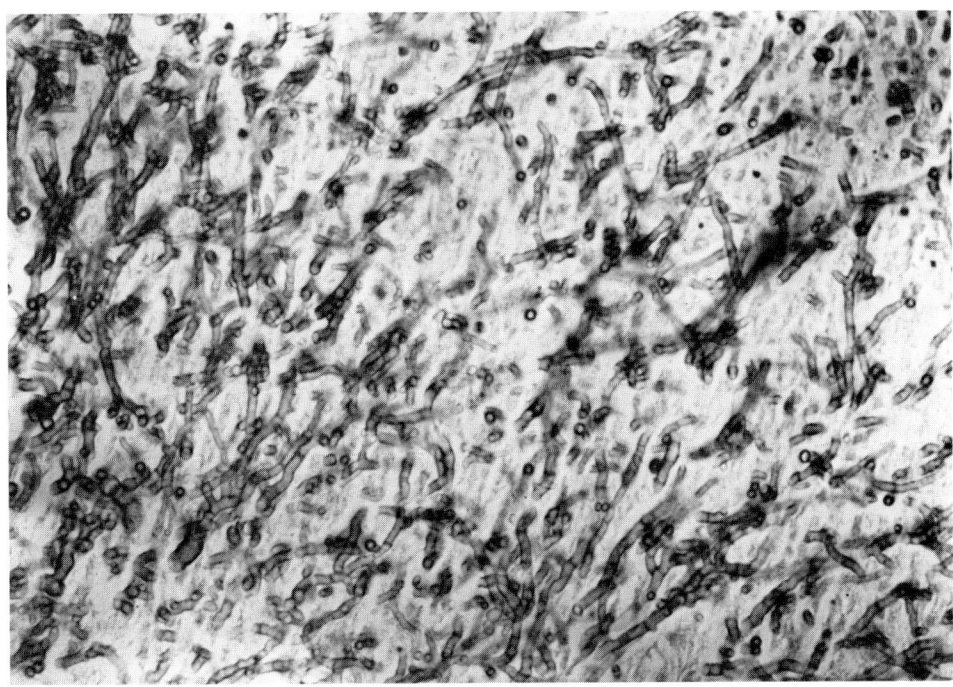

FIGURE 85. Conglomerate of fungal elements with septate branching hyphae resembling those of *Aspergillus* (Grocott, ×220).

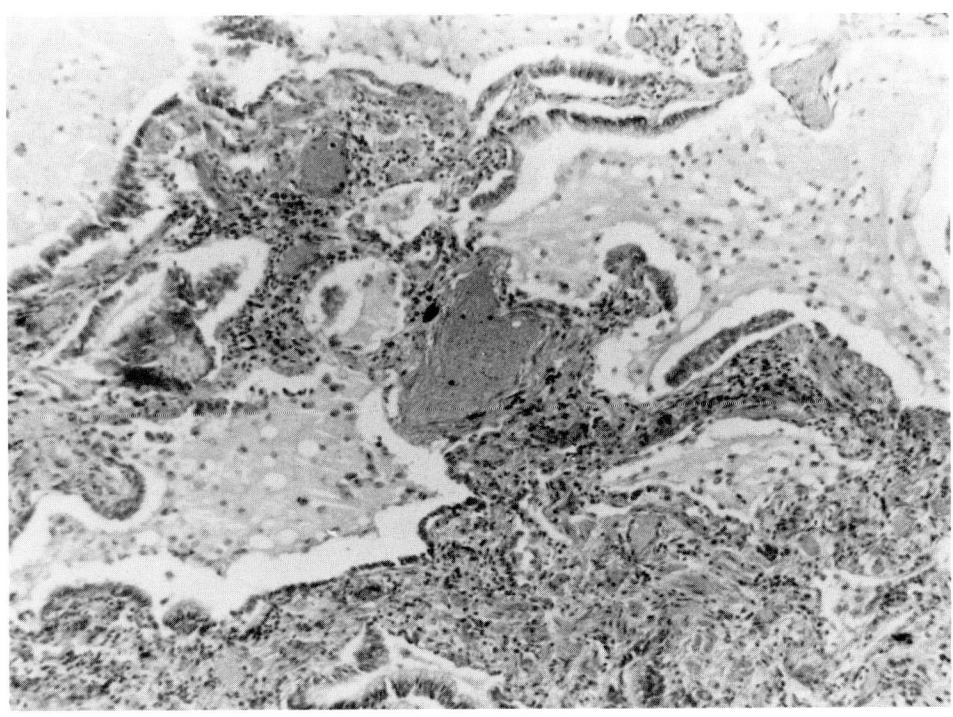

FIGURE 86. Interstitial lung fibrosis showing epithelization of some alveoli by proliferating bronchiolar epithelial cells and hyperplastic alveolar lining cells. The lumina contain proteinaceous substance and histiocytes (hematoxylin and eosin, ×130).

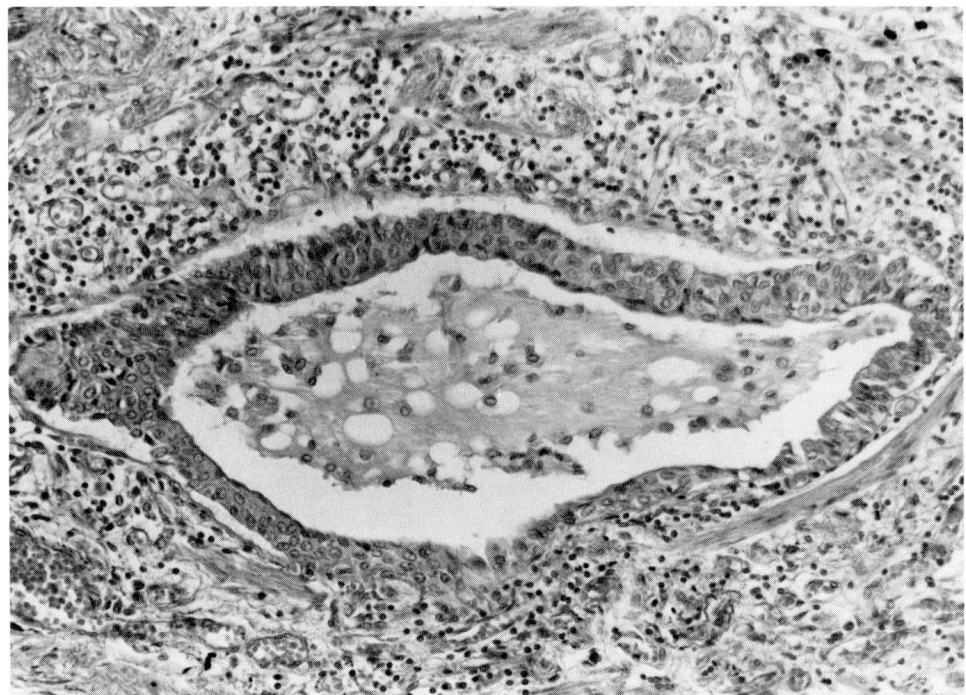

FIGURE 87. Squamous cell metaplasia of the bronchiolar mucosa (hematoxylin and eosin, ×175).

died of respiratory insufficiency before the radiation therapy had begun. The autopsy revealed lungs of firm consistency. On the cut surface, numerous large air spaces were visible, giving rise to a spongy pattern. No tumor, either in the lungs or in the lymph nodes, was found. A histological examination showed destruction of the normal acinar architecture of the lungs, replaced by fibrosis. The cells lining abnormal air spaces were flattened and in some areas cuboidal. Squamous cell metaplasis of the bronchioles was prominent (Figure 87). Bacteriological as well as serological investigations for precipitins to fungal antigens were negative. In a lung sample, AFB_1 at the concentration of 10 ng/g was detected by RIA. Following the same method for the determination of aflatoxin, no detectable levels were found in any of the lung samples of 10 age-matched control patients who had died from various causes (brain hemorrhage, myocardial infarct, renal insufficiency, chronic bronchitis, and chronic asthma).

B. COMMENT

The presence of aflatoxin found in lung tissue of five patients, two with lung cancer and three with pulmonary fibrosis, may be of particular relevance, indicating a possible association between lung damage and this carcinogen.

Although aflatoxin is primarily a hepatocarcinogen, this toxin has also been shown to produce neoplasms in extrahepatic tissues.[15] Aflatoxin, like many other carcinogens, requires metabolic activation for conversion to its ultimate carcinogenic form. Most of the research on aflatoxin metabolism has been performed using the liver as a source of metabolizing enzymes. Recent studies, however, have demonstrated that AFB_1 produces DNA adducts in respiratory cell cultures of humans[16,17] or explants of animals,[18,19] similar to those found in liver DNA *in vivo*.[20] Aflatoxicol, a potent mutagen and carcinogen, was found to be the major metabolite formed by tracheal explants from the Syrian hamster as well as from the rabbit.[18,19] The binding of AFB_1 to respiratory cell DNA has demonstrated the ability of respiratory tissue to activate

AFB_1 to the putative active intermediate, the AFB_1 2,3-oxide.[19] Since DNA adduct formation is a necessary condition for tumor induction, the data indicate that the airway exposure to AFB_1 presents a potential carcinogenesis risk to those exposed.

The first patient with lung alveolar cell carcinoma, bearing evidence of AFB_1 in his lung tissue, had a short-time exposure lasting only 3 months and developed symptoms 3 months later, ending in death within 11 months.

Considering that there is generally a long latent period between exposure to a carcinogen and the development of cancer in man, this finding is quite unusual, although similar cases of cancer deaths within a short time of initial exposure to aflatoxin were reported.[2,3]

Alveolar cell carcinoma of the lung is not a frequent form and its incidence has been estimated at between 3.9 and 5.0% of all lung cancers.[21] Its etiology is unclear. Focal congenital malformations of alveolar cells or inhalation of irritants, together with many other transient causes of lung damage, have been regarded as possible factors in tumor development. Chronic lung damage, however, has been commonly considered to play an important role in tumor genesis, while most of the tumors arise in relation to the areas of scarring in the lung.[21]

Disease in this patient was initially regarded as an inflammatory lung process of tuberculosis or of mycotic origin, and fibrous scars throughout the lungs, found at the autopsy, produced evidence of a preceding inflammation. Considering a possible mechanism of aflatoxin–related respiratory carcinogenesis, Baxter et al.[22] suggested that absorption of the carcinogen bound onto organic dust particles provides a potent inflammatory stimulus which enhances tissue susceptibility to the carcinogen. As the particles are deposited in the lungs, they retain the compound, allowing only its slow release before the particles are cleared from the respiratory tract. The tissue is therefore exposed to an effectively larger and more easily absorbed dose continuously over a long period of time.

Wang and Cerutti[23] studied the formation and removal of covalent AFB_1-DNA adducts under nontoxic conditions in epithelioid human lung cells and found that 60% of the total initial AFB_1 adducts were removed rapidly during the first 24 h of post-treatment incubation, while 40% of the adducts were removed only very slowly, leaving a sizable fraction of residual lesions in DNA over several generation times and suggested that persistent DNA lesions may represent a step forward in the initiation of malignant transformation.

Nonciliated bronchiolar epithelial (Clara) cells have been shown to be the target cells for actions of many toxicants in the lower airways of several mammalian species.[24] The susceptibility of these cells to toxicants is thought to be due to high levels of cytochrome P-450 monooxygenase, the most important enzyme system in xenobiotic metabolism.[24,25]

The relative potential for AFB_1 activation and susceptibility to AFB_1 toxicity of individual cell types has been studied in tracheal explants of the rabbit.[19] The results indicated that nonciliated tracheal epithelial cells (Clara cell "equivalent"), in the upper airways, appeared to be primarily responsible for carcinogen metabolism in this region of the respiratory tract.

Lung alveolar cell carcinoma is known to arise from epithelial cells in and distal to the terminal bronchioles including Clara cells.[21] It is therefore possible to suggest that the relatively short period between exposure to the carcinogen and cancer development in this patient may be related to the presence of these cells, which have been shown to be especially active in AFB_1 metabolism.

Aspergilloma producing aflatoxin in association with squamous cell carcinoma was found in the second patient, an agricultural worker was exposed to dust inhalation for several years. In his lung tissue, AFB_1 at a level as high as 2640 μg/kg was detected. It seems reasonable to suggest that an endogenic and prolonged production of the toxin was responsible for this extremely high level of the toxin found in lung tissue. The tumor was found to be limited only to the mycetoma cavity, that is to say, to the site particularly exposed to the carcinogen. This finding seems to provide strong evidence of the relationship between AFB_1 and cancer development.

The liver sample of this patient contained a lower concentration of AFB_1 (550 µg/kg) than that found in his lung. In animal experiments, it has been shown[17,26] that the respiratory system is capable of an early metabolization and activation of aflatoxin as well as of providing a site for rapid absorption prior to translocation to the liver.[26] Inhaled AFB_1-contaminated dust particles too large to reach the alveolar spaces may be transported via the mucociliary clearance to the esophagus and gastrointestinal tract.[7] The difference between AFB_1 levels found in the lung and those in the liver seems to support this hypothesis which was brought out by epidemiological evidence which indicated that workers exposed to AFB_1 via the respiratory route exhibited a greater incidence of tumors in a variety of sites, including the lungs, upper airways, and esophagus.[2]

Aflatoxin was found to be present in lung tissue of three patients, two agricultural workers and one textile worker, who died of interstitial pulmonary fibrosis associated with mycetoma in one of them.

It is known that people working with moldy hay, grain, or silage frequently fall ill with a lung disease known as farmer's lung.[27] However, it was generally assumed that this disease was caused by certain specific oversensitivity of some individuals toward antigens of fungi. Recently, a mycotoxicosis of the lung in agricultural workers who were exposed to a large amount of moldy silage via the respiratory tract has been reported by Emanuel et al.[28] The onset of the disease was acute and a chest roentgenogram of the patients revealed reticular and fine nodular densities scattered throughout the lungs comparable with interstitial pneumonia.

A histological examination of the lung biopsy sample showed a multifocal lung process which appeared related to the terminal bronchioles, alveoli, and interstitial areas. Immunological studies revealed no sensitivity to various fungal antigens. The authors pointed out that the clinical course as well as the negative immune reaction strikingly differentiated these patients from those with farmer's lung disease, and proposed that this disease was due to inhalation of fungal toxins from the environment. All their patients recovered and remained healthy, but carefully avoided exposure to moldy silage.

Similar etiopathogenic mechanisms could be considered in the three above-mentioned patients. Their illness began as an acute lung disease, clinically diagnosed as pneumonia or "influenza" pneumonitis. No precipitins to fungal antigens were proved in them, not even in the patient with lung mycetoma. All three patients still continued with their work after recovery, and thus reexposures could lead to a recurrent lung process resulting in fibrosis. Moreover, it could be assumed that the clearing ability of chronically damaged lungs is reduced, and the tissue may be therefore exposed to the carcinogen for a long time. Thus the risk of cancer development in such individuals may be high.

III. CONCLUSION

Although the present information relevant to the professional health risk from inhalation of aflatoxin-contaminated grain dust is rare, the potential risk seems apparent. It is already known that AFB_1 is carcinogenic toward the respiratory tract; AFB_1 can be metabolized by cultured human bronchus cells to provide an AFB_1-DNA adduct chromatographically identical to an adduct formed in the rat liver. The binding of AFB_1 to animal respiratory cells DNA demonstrated the ability of this tissue to activate AFB_1 to the putative intermediate, the AFB_1 2,3-oxide.

It is of critical importance that dust particles of a diameter enabling to reach the alveolar spaces contain the highest level of AFB_1, thus increasing a chance of exposure to and absorption of aflatoxin.

In addition, it is known that organic dust particles possess irritant properties which enhance tissue susceptibility to the carcinogen. Thus the inhaled particles may also act as a co-carcinogenic agent.

It is clear that present knowledge does not allow a definite conclusion of the cancer risk in men associated with respiratory exposure to aflatoxin and more research is needed in this area.

On the other hand, the recent data, though limited, indicate that the potential health hazard due to inhalation of airborne dust cannot be neglected.

REFERENCES

1. **Van Nieuwenhuize, J. P., Herber, R. F. M., de Bruin, A., Meyer, P. B., and Duba, W. C.,** Aflatoxinen: epidemiologisch onderzoek naar carcinogeniteit bij langurize "low level" exposite van een fabrickspopulatie, *T. Soc. Geneesk.,* 51, 754, 1973.
2. **Hayes, R. B., Van Nieuwenhuize, J. P., Raategever, J. W., and Ten Kate, F. J. W.,** Aflatoxin exposure in the industrial setting: an epidemiological study of mortality, *Food Chem. Toxicol.,* 22, 43, 1984.
3. **Deger, G. E.,** Aflatoxin — human colon carcinogenesis?, *Ann. Intern. Med.,* 85, 204, 1976.
4. **Dvořáčková, I.,** Aflatoxin inhalation and alveolar cell carcinoma, *Br. Med. J.,* 1, 691, 1976.
5. **Burg, W. R., Shotwell, O. L., and Saltzman, B. E.,** Measurements of airborne aflatoxins during the handling of contaminated corn, *Am. Ind. Hyg. Assoc. J.,* 42, 1, 1981.
6. **Burg, W. R., Shotwell, O. L., and Saltzman, B. E.,** Measurements of airborne aflatoxins during the handling of 1979 contaminated corn, *Am. Ind. Hyg. Assoc. J.,* 43, 580, 1982.
7. **Sorenson, W. G., Simpson, J. P., Peach, M. J., Thedell, T. D., and Olenchick, S. A.,** Aflatoxin in respirable corn dust particles, *J. Toxicol. Environ. Health,* 7, 669, 1981.
8. **Dashek, W. V., Eadie, T., Llewellyn, G. C., Olenchock, S. A., and Wirtz, G. H.,** Thin layer chromatographic analysis of possible aflatoxins within grain dusts, *J. Am. Oil Chem. Soc.,* 60, 563, 1983.
9. **Zennie, T. M.,** Identification of aflatoxin B_1 in grain elevator dusts in central Illinois, *J. Toxicol. Environ. Health,* 13, 589, 1984.
10. **Darke, C. S., Knowelden, J., Lacey, J., and Milford, A.,** Respiratory disease of workers harvesting grain, *Thorax,* 31, 294, 1976.
11. **Wicklow, D. T. and Shotwell, O. L.,** Intrafungal distribution of aflatoxins among conidia and sclerotia of *Aspergillus flavus* and *Aspergillus parasiticus, Can. J. Microbiol.,* 29, 1, 1982.
12. **Dvořáčková, I. and Polster, M.,** Relation between aflatoxin producing aspergilloma and lung carcinoma, *Microbiol. Alim. Nutr.,* 2, 187, 1984.
13. **Dvořáčková, I. and Píchová, D.,** Pulmonary interstitial fibrosis with evidence of aflatoxin B_1 in lung tissue, *J. Toxicol. Environ. Health,* 18, 153, 1986.
14. **Egmond, H. P., Paulsch, W. E., Sizov, E. A., and Schuller, P. L.,** Paper presented at the 4th IUPAC Conference on Mycotoxins, Lausanne, Switzerland, August 29 to 31, 1979.
15. **Wogan, G. N. and Newberne, P. M.,** Dose response characteristics of AFB_1 carcinogenesis, *Cancer Res.,* 27, 2370, 1967.
16. **Autrup, H., Essigman, J. M., Croy, R. G., Trump, B. F., Wogan, G. N., and Harris, C. C.,** Metabolism of aflatoxin B_1 and identification of the major aflatoxin B_1-DNA adducts formed in cultured human bronchus and colon, *Cancer Res.,* 39, 694, 1979.
17. **Stoner, G. D., Daniel, F. B., Schenck, K. M., Schut, H. A. J., Sandwisch, D. W., and Gohara, A. F.,** DNA binding and adduct formation of aflatoxin B_1 in cultured human and animal tracheobronchial and bladder tissues, *Carcinogenesis,* 3, 1345, 1982.
18. **Coulombe, R. A., Wilson, D. W., and Hsieh, D. P. H.,** Metabolism, DNA binding, and cytotoxicity of aflatoxin B_1 in tracheal explants from Syrian hamster, *Toxicology,* 32, 117, 1984.
19. **Coulombe, R. A., Jr., Wilson, D. W., Hsieh, D. P. H., Plopper, C. G., and Serabjit-Singh, C. J.,** Metabolism of aflatoxin B_1 in the upper airways of the rabbit: role of the non-ciliated tracheal epithelial cell, *Cancer Res.,* 46, 4091, 1986.
20. **Croy, R. G., Essigman, J. M., Reinhold, V. N., and Wogan, G. N.,** Identification of the principal aflatoxin B_1 DNA adduct formed *in vivo* in rat liver, *Proc. Natl. Acad. Sci. U.S.A.,* 75, 1745, 1978.
21. **Spencer, H.,** Bronchiolar-alveolar cell lung cancer, in *Pathology of the Lung,* Spencer, H., Ed., Pergamon Press, Oxford, 1985, 892.
22. **Baxter, C. S., Wey, H. E., and Burg, W. R.,** A prospective analysis of the potential risk associated with inhalation of aflatoxin-contaminated grain dusts, *Food Cosmet. Toxicol.,* 19, 765, 1981.
23. **Wang Tzu-Chien, V. and Cerutti, P. A.,** Formation and removal of aflatoxin B_1-induced DNA lesions in epitheloid human lung cells, *Cancer Res.,* 39, 5165, 1979.

24. **Boyd, M. R.,** Evidence for the Clara cell as a site of cytochrome P-450-dependent mixed function oxidase activity in lung, *Nature (London),* 269, 713, 1977.
25. **Serabjit-Singh, C. J., Wolf, C. R., Philpot, R. M., and Plopper, C. G.,** Cytochrome P-450: localization in the rabbit lung, *Science,* 207, 1469, 1980.
26. **Coulombe, R. A., Jr. and Sharma, R. P.,** Clearence and excretion of intratracheally and orally administered aflatoxin B_1 in the rat, *Food Chem. Toxicol.,* 23, 827, 1985.
27. **Spencer, H.,** Farmer's lung, in *Pathology of the Lung,* Spencer, H., Ed., Pergamon Press, Oxford, 497, 1985.
28. **Emanuel, W. A., Wenzel, B. S., and Lawton, B. R.,** Pulmonary mycotoxicosis, *Chest,* 67, 293, 1975.

INDEX

A

Acetyl-CoA dehydrogenase deficiency, 32
Acute aflatoxicosis, 11—12
Acute encephalopathy, 21
Acute hepatitis, 11
Acute liver dystrophy, 11
Adenocarcinoma, of colon, 12
Adenovirus 2, 109
Adenovirus 3, 31, 98—101, 109
Adenovirus 7, 31
Adenovirus 5, 109
Adenovirus 3/aflatoxin B_1 interactions in mouse, 97—107
    materials and methods, 97—98
    results, 98—107
Adrenal cortex, 53—54
Adrenocorticotropin (ACTH), 9
Aflatoxicol, 7—8
Aflatoxicosis, 11—12
Aflatoxin
    analysis of, 4
    biochemical effects of, 8—9
    carcinogenicity of, 5, 12—13, 113—129
    in Czechoslovakian liver cancer studies, 118
    effects of on man, 10—13
    factors modifying, 9—10
    in foodstuffs, 3
    hepatic concentrations of, 109
    history of, 2—3
    immunological effects of, 6—7, 115
    inhalation of, 135—146
        corn dust particles, 135—136
        in lung tissue, 136—146
    in liver cancer patients, 124—127
    metabolic activation of, 6—7
    mutagenicity of, 6
    nomenclature of, 2—8
    sources of, 3—4, 128
    teratogenic activity of, 5—6
    in third world liver carcinogenesis, 113—114
    toxicity of, 4—5
    viral interactions and, 30—31, 49—52, 94, 97, 103—105, 108—109, 115—116
Aflatoxin B, 2
Aflatoxin B_1 (AFB_1), 3—5, 60
    animal studies of, 65
    covalent binding of, 8
    ethnic differences in susceptibility to, 114
    in grain elevators, 136
    hepatic concentrations of, 108
    hepatotoxicity of, 34
    highly active forms of, 8
    in liver cancer, 126, 128
    mitochondrial enzyme activity and, 108
    mutagenicity of, 6
    neurotoxic effect of, 108
    relative cellular toxicity of, 147
    teratogenicity of, 6
    in Thai studies of Reye's syndrome, 34
    UV absorption spectra of, 56, 62
Aflatoxin B_2, 5, 61
Aflatoxin B_{2a}, 7
Aflatoxin B_1/influenza A virus interactions in mice, 94
Aflatoxin B_1/nitrosamine interactions, 116
Aflatoxin G, 2
Aflatoxin G_1, 5, 56
Aflatoxin G_2, 61
Aflatoxin G_{2a}, 7
Aflatoxin/HBV synergism and primary liver cancer, 115—116
Aflatoxin M, 2
Aflatoxin M_1 (AFM_1), 3, 5, 7, 60, 61, 128
Aflatoxin P_1 (AFP_1), 8, 128
Aflatoxin Q_1 (AFQ_1), 8
African studies, 114
Age differences in susceptibility, 4
Airborne dust, 135—136
Alcohol in liver cancer, 116—117, 124
Alcoholism, 117
Alimentary toxic aleukia (ATA), 1
Alveolar carcinoma, 12, 137—138, 147
Alzheimer's glia type II, 48—49
Anaplastic carcinoma, 135
Asian studies of Reye's syndrome, 29—30
Aspergillomas, 139—142, 147
Aspergillosis, 6
Aspergillus flavus, 2, 3, 34, 128—129, 136, 142
Aspergillus ochraceus, 2
Aspergillus parasiticus, 3
ATA, see Alimentary toxic aleukia

B

Bacillus brevis, 4
Bacillus megaterium, 4
Bacteriological findings in Reye's syndrome, 49—52
Balkan endemic nephropathy, 2
Barbiturate coma as treatment, 23
Beriberi, 2
Bile duct proliferation, 69
Biological tests, 4
Bisfurano-coumarin metabolites, 2—3
Bladder carcinomas, 5
Brain, 24—25, 27, 48
Breast feeding and aflatoxin exposure, 94
Bronchial carcinoma, 135, 146
Bursa Fabricii, 7

C

Cancer, see specific types
Cardiac beriberi, 2

Cardiac changes, 24
Carnitine deficiency, 32
Cecal coccidiosis, 6
Cellular immunity in Reye's syndrome, 31
Centers for Disease Control (CDC), 27
Cerebral edema, 23, 80, 88
Charcoal hyperperfusion, 23
Chemical analysis, 4
Chicken embryo bioassays, 4
Cholangiocarcinoma, 123—124
Choline, 9
Cirrhosis of liver, 12, 70, 116—117
Citreoviridin, 2
Clara cells, 147
Claviceps purpurea, 1
Colon cancer, 5, 12, 135
Corn dust particles, 135—136
Coxsackie virus, 31, 109
Craniectomy, 23
Crystallization in fat vacuoles of liver, 49
Czechoslovakian liver cancer studies, 117—129
    conclusion, 129
    materials and methods, 118
    results and comments, 118—129
        alcohol consumption, 124
        ELISA, 127—128
        histological investigation, 120
        HBV infection, 120, 124
        sources of aflatoxin, 128—129
Czechoslovakian National Influenza Surveillance, 109
Czechoslovakian Reye's syndrome studies, 43—112, see also Reye's syndrome

D

Dengue fever, 30
Diaphragm, 52, 87, 90, 91
Dicarboxylic aciduria, 26

E

Echo virus, 31
ELISA tests, 119, 127—128
Encephalopathy, 21, 34, 108
Endocrine status and carcinogenesis, 9
Enzyme-linked immunosorbate microassay, see ELISA tests
Epidemiologic data in Reye's syndrome, 27—30, 54
Exchange transfusion, 23
Exposure conditions, 10

F

Familial steatosis, 11, 93
Farmer's lung, 148
Fatty acids, 25, 34
Fatty liver, see Liver
Fetal growth retardation, 5—6
Fusarium sporotrichoides, 1

Fusarium poae, 1

G

Gallbladder cancer, 5, 135
Gastrointestinal cancer, 135
Genetic factors in Reye's syndrome, 31—32
Grain elevators, 136
Great Britain, Reye's syndrome in, 29—30
Gross pathology in Reye's syndrome, 23—24, 48
Guangxi (China), 115

H

HBV infection, 113—114, 117, 120
HeLa cells, 98
Hemangioendothelial carcinomas, 5
Hematologic findings in Reye's syndrome, 21
Hepatic cancer, see Liver cancer; Primary liver cancer
Hepatitis, 54, 59, 109, 114, 115
Herpes simplex, 31
Herpesvirus, 30, 109
High-performance liquid chromatography (HPLC), 4
Histological changes in Reye's syndrome, 24, 48—49
Hoeppli-Splendore phenomenon, 142
Hyaline membrane disease, 90—91
Hyperammonemia, 22, 26
Hypercatecholinemia, 27
Hypertyraminemia, 108
Hypoglycemia, 22, 27

I

Immunosuppression, 31, 35, 115
India, 11
Infectious hepatitis, 54
Inflammatory infiltration of liver, 49
Influenza A, 28, 59, 109
Influenza A(H_3N_2), 109
Influenza B, 28, 30, 109
Influenza HBsAg, 115
Interferon response, 35
Interstitial pancreatitis, 24
Interstitial pulmonary fibrosis, 144—145
Intrauterine exposure to aflatoxin, 94
Intrinsic and extrinsic toxins in Reye's syndrome, 32

J

Jamaican vomiting sickness, 32
Japanese studies of Reye's syndrome, 30
Juvenile cirrhosis, 12

K

Kenya, 11, 12
Khartuom, 12
Kidney, 24, 49, see also Renal tubules
Korean studies of Reye's syndrome, 30
Kuppfer cell activation, 50

Kwashiorkor, 11

L

Lipid metabolism alterations, 108
Lipid transport alterations, 108
Lipotrope deficiency, 9
Liver
 aflatoxin in, 109
 chronic damage to, 82
 diffuse fatty changes in, 48—50
 fatty changes in, 24, 69, 71, 84, 88
 inflammation in, 49
 nuclear granulation in, 73
 nuclear membrane changes in, 74—75
Liver biopsy, 24
Liver cancer, see also Primary liver cancer
 aflatoxin ingestion levels and, 13
 alcohol and, 116—117
 cirrhosis and, 116—117
 Czechoslovakian study of, 117—129
 mortality in, 114
Liver damage and Reye's syndrome, 66—85
Liver tumor incidence, 5
Lung alveoli, histologic changes in, 53
Lung tissue of exposed workers, aflatoxins in, 136—146, see also Aflatoxin, inhalation of
Lymphoid hyperplasia, 49

M

Macronodular cirrhosis, 116
Male/female differences in susceptibility, 4
Marck's disease, 6
Maxillary sinus cancer, 135
Measles, 30
Mesenteric lymph node necrosis, 52—53
Metabolic changes in Reye's syndrome, 25
Milk and dairy products, 3, 7
Mitochondrial injury, 32, 72
Mouse myocardium, 96
Mucinous adenocarcinoma of colon, 5
Muscle fibers in neonatal Reye's syndrome, 90
Mycotoxin porcine nepropathy, 2
Mycotoxins, 1—2
Myocardial steatosis, 49—51, 89, 96
Myxovirus, 30

N

National Reye's Syndrome Surveillance System (U.S.), 27
Neonatal Reye's syndrome, 85—94, see also Reye's syndrome, in neonates
Neurotransmitters and encephalopathy, 108
Nitroasamine/aflatoxin B_1 interactions, 116
Nonciliated brochiolar epithelial cells, 147
Nutritional factors and carcinogenesis, 9—10

O

Occupational risks and respirable aflatoxins, 135—149, see also Aflatoxin, inhalation of
Ochratoxin A, 2
Osteogenic carcinomas, 5

P

Pancreatitis, 24
Panreatic carcinoma, 5
Parainfluenza, 31
Paramyxovirus, 30
Penicillium citrevirides, 2
Penicillium viridicatum, 2
Phagocytosis impairment, 7
Philippines, 116
Polio vaccine and Reye's syndrome, 43
Primary liver cancer (PLC)
 aflatoxin levels and, 13
 geographical incidence of, 12
 histologic investigation of, 120
 incidence of, 119—120
 in third world countries, 113—116
 in Western countries, 116—117
Prostate cancer, 135
Protein deficiency, 9
Psoas muscle, 52, 87, 91, 93
Pulmonary alveoli in Reye's syndrome, 49, 53
Pulmonary surfactant insufficiency, 94

R

Radioimmunoassays (RIA), 4, 57
RDS, see Respiratory Distress Syndrome
Recurrent Reye's syndrome, 32
Renal pathologic changes, 24
Renal tubules, 50—51, 77—79, 89, 96
Reovirus, 31
Respirable aflatoxins and occupational risk, 135—149, see also Aflatoxin, inhalation of
Respiratiry Distress Syndrome (RDS), 85
Reticulum cell carcinoma, 135
Reye's syndrome (RS)
 aflatoxins and, 34, 54—69
  clinical and laboratory findings in, 57—59
  epidemiological data in, 59—60
  microscopic findings in, 60—61
  morphological findings in, 61—66
 age and sex distribution of, 59
 bacteriological, virological, and toxicological findings in, 49—52
 cerebral spinal fluid (CSF), 22
 with chronic liver damage, 66, 80—85
 clinical and laboratory data on, 43—47
 clinical symptoms of, 21, 44
 Czechoslovakian cases of 1958—1970, 43—54
 Czechoslovakian follow-up study of, 43—112

bacteriological, virological and toxicological findings, 47—54
chronic liver damage, 66, 80—84
clinical and laboratory data, 43—46
epidemiological data, 54
experimental models, 94—109
morphological features, 46—47
in newborns, 84, 90—94
relationship with aflatoxin, 54—66
epidemiologic data in, 27—30, 54
etiology of, 30
experimental models of, 94
gross pathology in, 23—24, 48
histologic changes in, 24, 28—35, 43—49
laboratory evaluation of, 21—22, 47
metabolic abnormalities in, 25
month of onset of, 59
morphological features of, 23—25, 48
mouse model of, 94—107
in neonates, 85—94
clinical data and, 87
degree of fatty change in, 92
familial data and, 87
ornithine transcarbamylase (OTC) deficiency, 26
pathogenesis of, 25—27
recurrent, 32
rural occurrence of, 59
therapy in, 22—23
toxins in, 32
viruses associated with, 30—31
RIA, see Radioimmunoassays

S

Salicylates and Reye's syndrome, 32—33, 109
Salmonellosis, 6
San Diego Medical Center treatment protocol, 22—23
Senegal, 114
Serotological findings in Reye's syndrome, 21—22
Sex differences in resistance, 9
Sex hormone effects, 9
SIDS, see Sudden infant death syndrome
Smallpox vaccination and Reye's syndrome, 43
Southeast Asian studies, 114
Spectrophotometry, 55—57
Spleen, 52—53

Squamous cell carcinoma, 5, 141
"Stachybotriotoxicosis", 1
Stachybotrys atra, 1
St. Anthony's fire, 1
Sudden infant death syndrome (SIDS), 85
Swaziland, 12, 113

T

Taiwanese studies, 114
Thailand, 12, 34, 113—114
Thin-layer chromatography (TLC), 4, 55—57
Thymus gland regression, 7
Tissue cultured cells, 4—5
Tongue carcinomas, 5
Toxicological findings in Reye's syndrome, 49—52
Toxin/toxin interactions, 34—35
Trabecular carcinoma of liver, 120—121
Transplacental transmission, 94
Turkey X disease, 2, 5

U

Uganda, 12, 113, 114
Ultrastructural changes in Reye's syndrome, 24—25
Undorn encephalopathy, 34
Urinary bladder, see Bladder

V

Vaccination and Reye's syndrome, 43
Varicella virus, 31
Vegetable foods, 3
Viral hepatitis, 115
Viral infection and Reye's syndrome, 30—31
Virus/aflatoxin interactions, 34—35, 103—105, 108—109
Vitamin B_{12}, 9—10

Y

Yellow rice disease, 2

Z

Zambia, 114
Zona fasciculata, 49

DATE DUE